高等职业教育计算机类专业新型一体化教材

Android Studio 项目开发教程
——从基础入门到乐享开发

彭 艳 主 编
孙宏伟 杨 欧 罗齐熙 副主编

电子工业出版社
Publishing House of Electronics Industry
北京·BEIJING

内 容 简 介

本书是以 Android 应用开发为基础的从入门到精通的实践项目教材，采用任务驱动的方式，将 Android 开发中的每一个具体环节都融入具有代表性的企业一线项目实践，以实现通过完成项目掌握能力的目的，最终完成整个 Android APP。本书中所有的知识点均有理论解析和实际应用，选择的"购物商城 APP"项目来源于企业一线，该项目严格贯彻执行行业的开发流程和规范。本书既可作为高职院校计算机应用技术、移动互联技术和物联网技术等专业的教材，还可作为对 Android 应用开发感兴趣的读者的参考书。

未经许可，不得以任何方式复制或抄袭本书之部分或全部内容。
版权所有，侵权必究。

图书在版编目（CIP）数据

Android Studio 项目开发教程：从基础入门到乐享开发 / 彭艳主编. 一北京：电子工业出版社，2020.7（2023.8 重印）
ISBN 978-7-121-37505-7

Ⅰ. ①A… Ⅱ. ①彭… Ⅲ. ①移动终端—应用程序—程序设计—高等学校—教材 Ⅳ. ①TN929.53

中国版本图书馆 CIP 数据核字（2019）第 216486 号

责任编辑：李　静　　　　　　文字编辑：张　慧
印　　刷：北京盛通商印快线网络科技有限公司
装　　订：北京盛通商印快线网络科技有限公司
出版发行：电子工业出版社
　　　　　北京市海淀区万寿路 173 信箱　邮编：100036
开　　本：787×1 092　1/16　印张：10.75　字数：276 千字
版　　次：2020 年 7 月第 1 版
印　　次：2023 年 8 月第 6 次印刷
定　　价：35.80 元

凡所购买电子工业出版社图书有缺损问题，请向购买书店调换。若书店售缺，请与本社发行部联系，联系及邮购电话：(010) 88254888，88258888。
质量投诉请发邮件至 zlts@phei.com.cn，盗版侵权举报请发邮件至 dbqq@phei.com.cn。
本书咨询联系方式：(010) 88254604，lijing@phei.com.cn。

前 言

Android 是 Google 公司开发的基于 Linux 的开源操作系统，主要应用于智能手机、平板电脑等移动设备。随着物联网技术、移动互联技术和人工智能技术的蓬勃发展，Android 系统除在智能手机领域外，在智能物联、智能硬件、穿戴设备等领域也得到了大规模推广。随着 Android 的迅速发展，各领域对 Android 开发人才的需求猛增，因此越来越多的人开始学习 Android 应用开发，以适应市场需求，寻求更广阔的发展空间。

本书采用全新的开发工具 Android Studio，并从初学者的角度出发，围绕一个实际的"购物商城 APP"项目的开发过程，精心组织和编排了知识体系和实践环节。全书共分 9 章，内容涵盖界面设计、UI 交互、页面跳转、数据存储、数据库编程、网络编程、多媒体编程等实际开发中不可或缺的知识和技能。本书由浅入深地讲解每个知识点，每个实际制作环节都配有详细的步骤说明和代码。

众所周知，Android 开发主要使用的是 Java 语言。因此，在学习本书之前，需要具备 Java 语言及面向对象程序设计的基础知识。建议初学者从头开始，循序渐进地学习，并且反复练习书中的案例，以做到熟能生巧、融会贯通；有一定基础的编程人员，可在本书中选择自己感兴趣的章节进行跳跃式学习，但最好动手实践本书中的全部案例。

本书是新型一体化教材，配备丰富的教学资源，包括 PPT、素材、源码、微课视频等，可联系 QQ：1096074593 索取，或者通过封面二维码直接扫描下载。

在编写本书的过程中，深圳职业技术学院的多位领导、老师为本书提出了非常宝贵的建议，在此特别感谢他们的帮助。

彭 艳
2020 年 4 月

目 录

第 1 章 Android 入门 ·· 1
1.1 Android 简介 ·· 1
1.2 Android Studio 开发环境搭建 ··· 3
1.2.1 Android Studio 介绍 ··· 3
1.2.2 开发环境配置要求 ··· 3
1.2.3 搭建开发环境 ··· 3
1.3 开发 Android 应用程序 ··· 8
1.3.1 新建 Android 应用程序 ·· 8
1.3.2 Android 项目结构 ··· 11
1.3.3 创建 Android 模拟器 ··· 15
1.3.4 在模拟器上运行 APP ··· 18
1.3.5 连接手机运行 APP ·· 19
1.4 认识项目——购物商城 APP ··· 22
1.4.1 开发背景 ·· 22
1.4.2 系统功能设计 ·· 23
1.4.3 项目包结构说明 ··· 23
1.4.4 系统预览 ·· 24
1.5 本章小结 ·· 24
1.6 本章习题 ·· 24

第 2 章 Android 用户界面设计 ·· 25
2.1 UI 设计的相关概念 ·· 25
2.1.1 View ·· 25
2.1.2 ViewGroup ·· 26
2.2 控制 UI 界面 ·· 27
2.2.1 使用 XML 布局文件控制 UI 界面 ·· 27
2.2.2 开发自定义的 View 类 ··· 28
2.3 布局管理器 ··· 29
2.3.1 相对布局管理器 ··· 29

	2.3.2 线性布局管理器	31
	2.3.3 帧布局管理器	32
	2.3.4 表格布局管理器	33
	2.3.5 网格布局管理器	34
	2.3.6 布局管理器的嵌套	34
2.4	购物商城 APP 的布局设计	35
	2.4.1 购物商城 APP 商城首页布局	35
	2.4.2 个人中心页面布局	42
2.5	本章小结	45
2.6	本章习题	45

第 3 章 常用 UI 组件 46

3.1	常用组件	46
	3.1.1 文本类组件	46
	3.1.2 按钮类组件	48
	3.1.3 图像类组件	55
3.2	常见对话框	57
	3.2.1 通过 Toast 类显示消息提示框	57
	3.2.2 使用 AlertDialog 类实现对话框	57
	3.2.3 使用 Notification 类在状态栏上显示通知	59
3.3	购物商城 APP 的 UI 交互	59
	3.3.1 商城首页底部的页面选择	59
	3.3.2 用户登录	61
	3.3.3 用户注册	65
3.4	本章小结	68
3.5	本章习题	69

第 4 章 基本程序单元 Activity 70

4.1	Activity 概述	70
4.2	创建、配置、启动和关闭 Activity	72
	4.2.1 创建 Activity	72
	4.2.2 配置 Activity	73
	4.2.3 启动和关闭 Activity	73
	4.2.4 Intent 介绍	74
	4.2.5 显式 Intent 和隐式 Intent	74
4.3	多个 Activity 的使用	76
	4.3.1 使用 Bundle 在 Activity 之间交换数据	76
	4.3.2 调用另一个 Activity 并返回结果	76
4.4	使用 Fragment	77
	4.4.1 Fragment 的生命周期	77

4.4.2　创建 Fragment ……………………………………………………… 78
　　　4.4.3　在 Activity 中添加 Fragment ……………………………………… 79
　4.5　购物商城 APP 页面的跳转和数据传递 ………………………………………… 80
　　　4.5.1　商城底部的页面切换 ……………………………………………… 80
　　　4.5.2　个人中心页面—登录页面—注册页面的跳转 …………………… 86
　　　4.5.3　登录后跳转至个人中心页面 ……………………………………… 88
　4.6　本章小结 …………………………………………………………………………… 89
　4.7　本章习题 …………………………………………………………………………… 90

第5章　数据存储技术 ……………………………………………………………………… 91

　5.1　SharedPreferences 存储 …………………………………………………………… 91
　　　5.1.1　获取 SharedPreferences 对象 ……………………………………… 92
　　　5.1.2　向 SharedPreferences 文件存储数据 ……………………………… 92
　　　5.1.3　读取 SharedPreferences 文件中存储的数据 ……………………… 92
　5.2　文件存储 …………………………………………………………………………… 94
　　　5.2.1　内部存储 …………………………………………………………… 94
　　　5.2.2　外部存储 …………………………………………………………… 95
　5.3　购物商城 APP 的信息存储 ……………………………………………………… 97
　　　5.3.1　用户注册信息的存储 ……………………………………………… 97
　　　5.3.2　免验证快速登录功能 ……………………………………………… 98
　　　5.3.3　退出后清除 SharedPreferences …………………………………… 102
　5.4　本章小结 …………………………………………………………………………… 103
　5.5　本章习题 …………………………………………………………………………… 103

第6章　数据库编程 ………………………………………………………………………… 104

　6.1　SQLite 数据库简介 ……………………………………………………………… 104
　6.2　创建数据库 ………………………………………………………………………… 104
　6.3　SQLite 数据库的操作 …………………………………………………………… 105
　6.4　数据信息显示控件 ………………………………………………………………… 106
　　　6.4.1　ListView 介绍 ……………………………………………………… 106
　　　6.4.2　RecyclerView 介绍 ………………………………………………… 107
　6.5　购物商城 APP 的数据库编程 …………………………………………………… 107
　　　6.5.1　购物商城 APP 的数据库设计 ……………………………………… 108
　　　6.5.2　商品分类模块 ……………………………………………………… 117
　　　6.5.3　购物车模块 ………………………………………………………… 123
　6.6　本章小结 …………………………………………………………………………… 124
　6.7　本章习题 …………………………………………………………………………… 124

第7章　网络编程 …………………………………………………………………………… 125

　7.1　通过 HTTP 访问网络 …………………………………………………………… 125
　　　7.1.1　发送 GET 请求 ……………………………………………………… 126

7.1.2　发送 POST 请求 ·· 126
　7.2　解析 JSON 格式数据 ·· 126
　　　7.2.1　JSON 简介 ··· 126
　　　7.2.2　解析 JSON 数据 ·· 127
　7.3　网络查询手机号码归属地 ··· 128
　7.4　本章小结 ··· 132
　7.5　本章习题 ··· 132

第 8 章　多媒体编程 ··· 133

　8.1　动画 ·· 133
　　　8.1.1　补间动画 ··· 133
　　　8.1.2　逐帧动画 ··· 135
　8.2　音频与视频 ·· 137
　　　8.2.1　使用 MediaPlayer 类播放音频 ······························· 137
　　　8.2.2　使用 SoundPool 类播放视频 ·································· 139
　　　8.2.3　使用 VideoView 组件播放视频 ······························· 140
　8.3　商品详情页面的背景音乐 ··· 140
　8.4　本章小结 ··· 143
　8.5　本章习题 ··· 143

第 9 章　实现购物商城 APP 的其他功能 ······························ 144

　9.1　用户身份验证与注册 ··· 144
　9.2　添加商品到购物车 ··· 145
　　　9.2.1　显示商品详细信息 ·· 145
　　　9.2.2　将商品加入购物车 ·· 149
　　　9.2.3　查看、编辑购物车 ·· 151
　　　9.2.4　购物车结算 ·· 156

附录 A　素材说明 ··· 158

 # 第 1 章　Android 入门

 ## 1.1　Android 简介

　　Android 平台由操作系统、中间件、用户界面和应用软件组成，是一个真正开放的移动开发平台。Android 公司最初并不是由谷歌公司创办的，而是由 Andy Rubin 于 2003 年 10 月创办，Andy 后来被称为 Android 之父。谷歌公司于 2005 年收购了 Andy 创建的 Android 公司，于 2007 年对外展示了名称为 Android 的移动操作系统，并且宣布建立一个全球性的联盟组织。该组织由 34 家手机制造商、软件开发商、电信运营商及芯片制造商共同组成，并与 84 家硬件制造商、软件开发商及电信运营商组成手持设备联盟（Open Handset Alliance）来共同研发和改良 Android 系统。

　　Android 系统是基于 Linux 系统内核开发设计的，谷歌公司在该内核之上开发了自己的 Dalvik Java 虚拟机。因为采用 Java 虚拟机，所以在 Android 系统的平台上开发原生 APP 所用的开发语言是 Java。到目前为止，Android 系统已经是全球最大的智能手机操作系统，最新版本是 Android 8.0。

　　Android 系统的体系架构如图 1.1 所示。

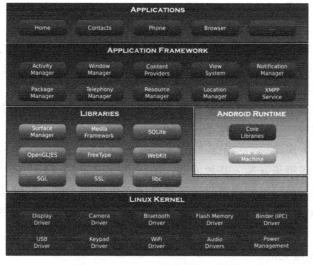

图 1.1　Android 系统的体系架构

从图 1.1 中可以看出 Android 系统大致可以分为 Linux 内核层（Linux Kernel）、系统运行层、应用框架层（Application Framework）、应用层（Applications）四层架构。

（1）Linux 内核层。

Android 系统是基于 Linux 2.6 内核的，Linux 内核层为 Android 设备的各种硬件提供了底层的驱动，如显示驱动、音频驱动、照相机驱动、蓝牙驱动、WiFi 驱动、电源管理驱动等。

（2）系统运行层。

系统运行层通过一些 C/C++库为 Android 系统提供主要的特性支持，如 SQLite 库提供了数据库的支持，OpenGL|ES 库提供了 3D 绘图的支持，Webkit 库提供浏览器内核的支持等。同时，在这一层还有 Android 系统运行时需要的库，如一些核心库能够允许开发者使用 Java 来编写 Android 应用。其中的关键是 Dalvik 虚拟机，它使得每个 Android 应用都能够运行在独立的进程中，并且拥有一个自己的 Dalvik 虚拟机实例，相比 Java 虚拟机（JVM），Dalvik 虚拟机是专门为移动设备定制的，它对手机内存、CPU 性能不佳等情况做了优化处理。

（3）应用框架层。

应用框架层主要提供了构建应用时可能用到的 API，Android 系统自带的一些核心应用程序就是使用这些 API 完成的，开发者可以通过使用这些 API 构建自己的应用程序，如活动管理器、视图系统、内容提供器、通知管理器等。

（4）应用层。

所有安装在手机上的应用程序都属于应用层，如系统自带的联系人、短信等程序，或者从 Google Play 上下载的程序，包括开发者自己开发的应用程序。

以下介绍 Android 系统的优势及特性。

（1）开源。

Android 系统完全开源，由于本身的内核是基于开源的 Linux 系统内核的，所以 Android 系统从底层系统到上层用户类库、界面等都是完全开放的。任何个人和组织都可以查看和学习源代码，也可以基于谷歌公司发布的版本制作自己的系统。华为、小米、三星等手机厂商都拥有自己个性化的 Android 系统。相对于谷歌公司发布的 Android 系统版本，很多手机厂商为突出自己的优势，在一些功能上做了优化。

（2）多元化设备支持。

Android 系统除在智能手机上被应用外，还在平板电脑、互联网电视、车载导航仪、智能手表及一些其他智能硬件上被广泛应用，如小米的平板电脑、电视，华为的车载导航仪、手表等。另外，围绕自动驾驶相关的系统也是利用 Android 系统进行开发的。

（3）Dalvik 虚拟机。

Dalvik 虚拟机相对于 Sun VM 来说有很多不同。Dalvik 在低速 CPU 上表现的性能更高，对内存的使用也更高效，这恰恰是移动设备所需要的。

（4）开放的第三方应用。

因为谷歌公司一贯秉承开源、开放原则，所以在 Android 系统上开发 APP、发布 APP 要相对容易一些。开发人员可以根据自己的需要调用手机中的 GPS、陀螺仪、摄像头等硬件设备，也可以访问本地联系人、日历等信息，并进行拨打电话、发送短信等操作。在 Android 系统上开发应用也不需要谷歌公司认证，所以 Android 系统的应用市场比较丰富。

（5）与 Google 系统无缝结合。

Android 系统可以和 Google 系统的地图服务、邮件系统、搜索服务等无缝结合。

1.2　Android Studio 开发环境搭建

1.2.1　Android Studio 介绍

Android Studio 是谷歌公司针对 Android 系统开发推出的，基于 IntelliJ IDEA 的 Android 应用开发集成开发环境（IDE），涵盖了所有 Android 应用开发相关的功能。越来越多的项目开发选择 Android Studio 而不再固守 Eclipse，下面介绍 Android Studio 的主要特点。

（1）Android Stuido 由谷歌公司推出。毫无疑问，这是它最大的优势，它是谷歌公司大力支持的一款基于 IntelliJ IDEA 的改造的 IDE。

（2）Android Stuido 在启动速度、响应速度、内存占用等性能上都优于 Eclipse。

（3）UI 界面更漂亮。Android Stuido 自带的 Darcula 黑色主题非常美观。相比而言，Eclipse 的主题略显单调。

（4）更加智能。Eclipse 代码的输入补齐和修正提示对于开发者来说意义重大，而 Android Studio 则更加智能，它增加了智能保存，开发者熟悉 Android Studio 后其工作效率会大大提升。

（5）整合了 Gradle 构建工具。Gradle 是一个新的构建工具，可以说，Gradle 集合了 Ant 和 Maven 的优点，无论是配置、编译，还是打包都非常方便。

（6）强大的 UI 编辑器。Android Studio 的编辑器非常智能，除吸收 Eclipse 的优点外，还自带了多设备的实时预览，有利于界面开发。

（7）内置终端。Android Studio 内置终端，这对于习惯命令行操作的开发者来说非常方便。

（8）更完善的插件系统。Android Studio 支持各种插件，如 Git、Markdown、Gradle 等。

（9）完美整合版本控制系统。Android Studio 在安装时就自带如 GitHub、Git、SVN 等流行版本的控制系统，方便项目迭代更新。

1.2.2　开发环境配置要求

Android Studio 开发环境配置要求如下。

（1）系统版本：Windows 8（64 位）/Windows 10（64 位）。

（2）计算机内存：推荐 4GB。

（3）Java SDK 版本：jdk-8u171-windows-x64（64 位）。

（4）Android Studio 版本：android-studio-ide-173.4720617-windows。

1.2.3　搭建开发环境

1. 安装 JDK（Java Development Kit）

在 JDK 官网 http://www.oracle.com/technetwork/java/javase/downloads/index.html 下载安装包，成功下载安装包后，进行安装。安装步骤如下。

(1)双击 JDK 安装包,根据提示选择安装目录,单击"下一步"按钮,完成安装。
(2)继续安装 JRE,安装时把 JRE 的安装目录更改为与 JDK 相同的安装目录。

注:若无安装目录要求,则无须做任何修改,按照默认设置单击"下一步"按钮,直至安装完毕。成功安装 JDK 后,设置环境变量,操作如下。

① 单击"控制面板"→"系统"→"高级系统设置"→"高级"→"环境变量"→"系统变量"选项。

② 在弹出的对话框中设置"变量名"为 JAVA_HOME,设置"变量值"为 JDK 的安装目录,如图 1.2 所示,单击"确定"按钮。

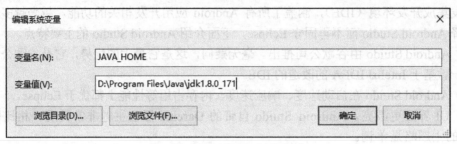

图 1.2　设置变量名和变量值

③ 如果 PATH 属性已经存在,则可以直接编辑环境变量,即在原来的内容后追加";%JAVA_HOME%\bin",如图 1.3 所示。

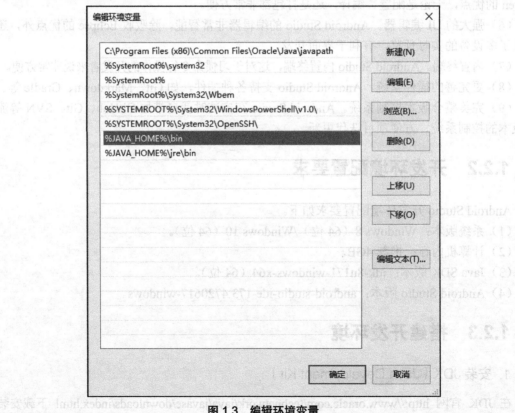

图 1.3　编辑环境变量

④单击"新建"按钮,在弹出的对话框中设置"变量名"为 CLASSPATH,设置"变量值"如图 1.4 所示。

图 1.4 设置变量名及变量值

⑤检查环境变量的配置是否成功。单击"开始"→"运行",输入"cmd",弹出命令运行窗口,输入"JAVAC",如输出帮助信息则为配置正确,如图 1.5 所示。

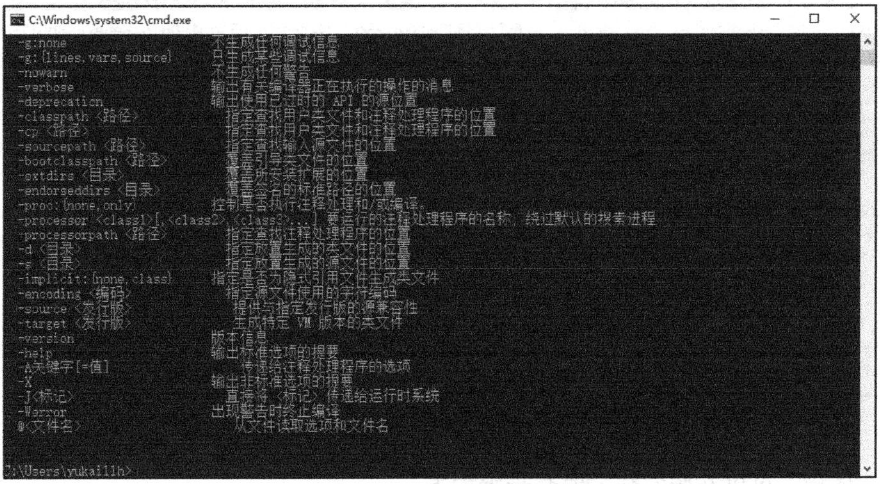

图 1.5 检查环境变量的配置是否成功

⑥查看 Java 版本信息。输入"Java-version"并按回车键,出现如图 1.6 所示的信息,即为安装成功。

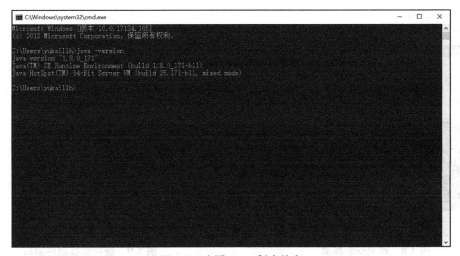

图 1.6 查看 Java 版本信息

2. 安装 Android Studio

在 Android Studio 的官方网站中，可以很方便地下载 Android Studio，访问 http://www.android-studio.org/，根据当前操作系统选择对应的版本，并进行下载，官网下载页面如图 1.7 所示。

图 1.7 Android Studio 官网下载页面

下载完成后，可以开始安装，安装步骤如下。

（1）直接点击安装包，将会弹出"打开文件-安全警告"对话框，单击"运行"按钮，开始安装，安装完成后的欢迎安装界面如图 1.8 所示。

（2）单击"Next"按钮，弹出选择安装组件窗口，如图 1.9 所示。采用默认设置，继续单击"Next"按钮。

图 1.8 欢迎安装界面

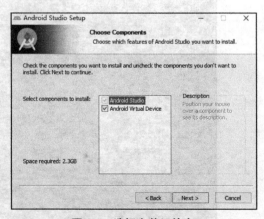

图 1.9 选择安装组件窗口

（3）在如图 1.10 所示的配置安装路径窗口中，指定 Android Studio 的安装路径。单击"Next"按钮，打开选择快捷方式菜单文件夹窗口，选择将 Android Studio 的快捷方式创建在开始菜单中的指定文件夹下，这里采用默认配置，如图 1.11 所示。

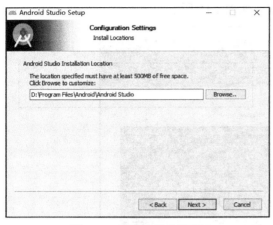 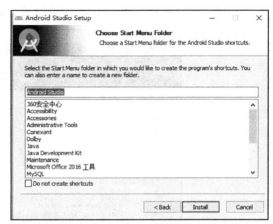

图 1.10　配置安装路径窗口　　　　　　图 1.11　选择快捷方式菜单文件夹窗口

（4）单击"Install"按钮，开始安装，等待安装完成即可。

3. 启动 Android Studio

（1）启动 Android Studio，弹出如图 1.12 所示的导入配置对话框。初次使用时，可采用默认配置。

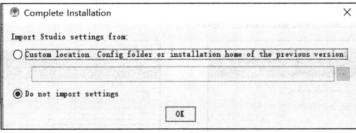

图 1.12　导入配置对话框

（2）单击"OK"按钮，进入 Android Studio，弹出如图 1.13 所示的启动加载框。加载完成后弹出如图 1.14 所示的对话框，询问是否设置代理，如果存在有效的代理地址，则可以单击"Setup Proxy"按钮，添加代理地址，否则单击"Cancel"按钮。

图 1.13　启动加载框　　　　　　　　图 1.14　设置代理对话框

（3）继续单击，直到出现选择安装类型对话框。这里包含两个选项，"Standard"和"Custom"，即标准安装和自定义安装，如果在本机中尚未安装 Android SDK，则建议直接选择标准安装。

（4）单击"Next"按钮，进入主题设置页面，如图 1.15 所示。

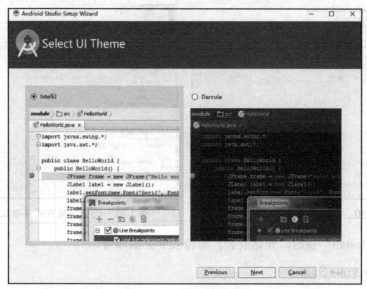

图 1.15　主题设置界面

（5）设置完主题后，开始安装 SDK，SDK 信息展示界面如图 1.16 所示。此时仅需等待安装完成即可。下载 SDK 界面如图 1.17 所示。

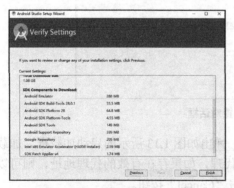　　　　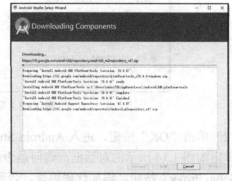

图 1.16　SDK 信息展示界面　　　　　　　　图 1.17　下载 SDK 界面

1.3　开发 Android 应用程序

1.3.1　新建 Android 应用程序

Android Studio 安装完成后，如果还没有创建项目，则弹出欢迎对话框。在该对话框中可以创建新项目，打开已经存在的项目，导入项目等。在 Android Studio 中，一个项目（project）

相当于一个工作空间，一个工作空间中可以包含多个模块（Module），每个 Module 对应一个 Android 应用。下面介绍如何创建第一个 Android 应用。

（1）在 Android Studio 的欢迎对话框中，单击"Start a new Android Studio project"按钮，进入"Create New Project"（创建新项目）对话框中，在"Application name"文本框中输入应用程序名称"Hello World"；在"Company domain"文本框中输入公司域名；在"Project location"文本框中输入项目保存的位置，如图 1.18 所示。

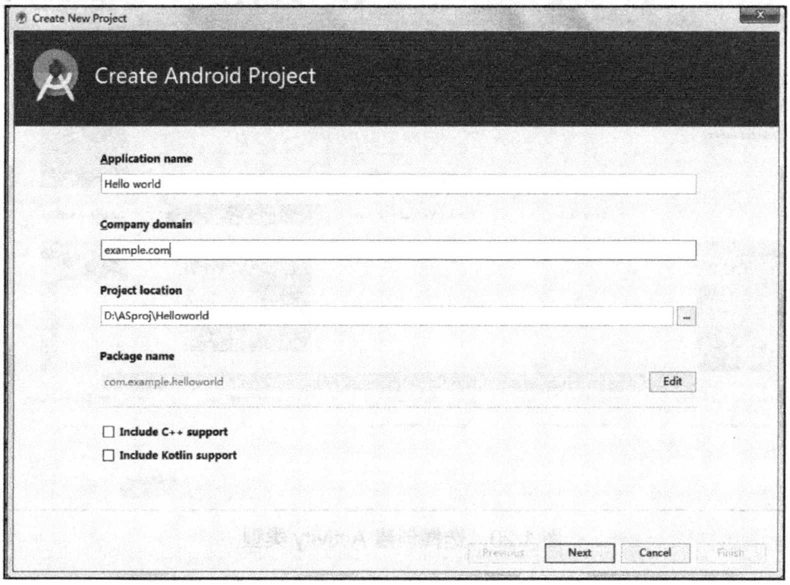

图 1.18　创建新项目

（2）单击"Next"按钮，进入"Target Android Devices"（选择目标设备）对话框。在该对话框中，首先选择"Phone and Tablet"复选框，然后在"Minimum SDK"（最低 SDK 版本）下拉列表框中选择默认的"API 15: Android 4.0.3（IceCreamSandwich）"，如图 1.19 所示。

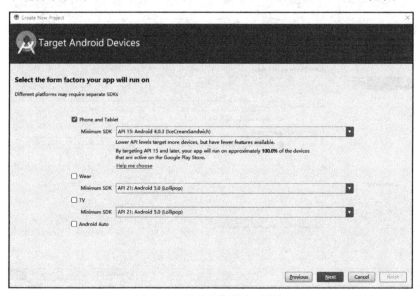

图 1.19　选择目标设备对话框

（3）单击"Next"按钮，进入"Add an Activity to Mobile"（选择创建 Activity 类型）对话框，在该对话框中，系统将列出一些用于创建 Activity 的模板。此时，既可根据需要进行选择，也可以选择不创建 Activity（选择 Add No Activity）。这里选择创建一个空白的 Activity，即 Empty Activity，如图 1.20 所示。

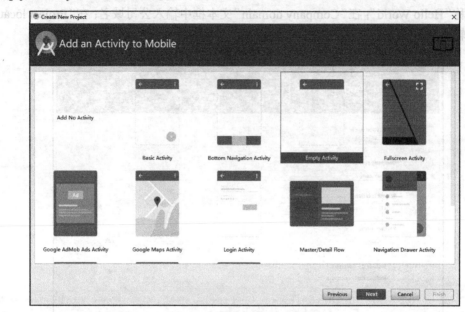

图 1.20　选择创建 Activity 类型

（4）单击"Next"按钮，进入"Customize the Activity"（自定义 Activity）对话框，在该对话框中，可以设置自动创建的 Activity 的类型名和布局文件名称，这里采用默认配置，如图 1.21 所示。

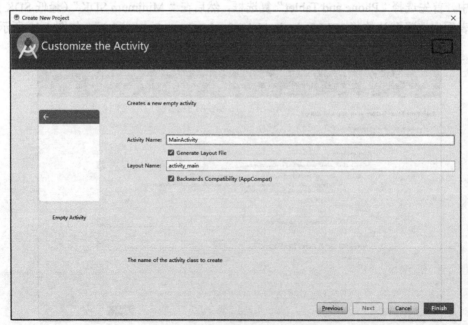

图 1.21　自定义 Activity

（5）单击"Finish"按钮，将弹出创建进度对话框，创建完成后，该对话框将自动消失，并同时打开该目录。

（6）默认情况下，启动项目时会弹出小贴士对话框，单击"Close"按钮，即可进入 Android Studio 的主页，同时打开已创建好的项目，默认显示 MainActivity.java 文件的内容，选择 activity_main.xml 选项卡，显示布局编辑器。Android Studio 的主页如图 1.22 所示。

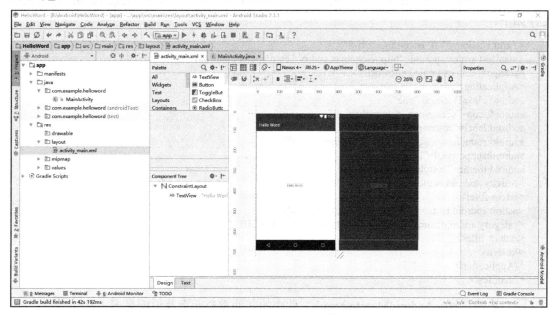

图 1.22　Android Studio 的主页

1.3.2　Android 项目结构

默认情况下，在 Android Studio 中创建 Android 项目后，将默认生成如图 1.23 所示的项目结构。在 Android Studio 中，提供了多种项目结构类型。其中，最常用的是 Android 项目结构类型和 Project 项目结构类型。项目创建完成后，默认采用 Android 项目结构类型。

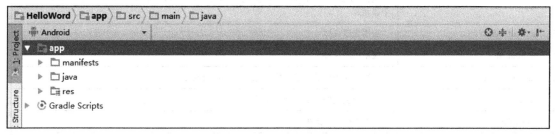

图 1.23　Android 项目结构

在图 1.23 中可看出，Android 项目主要包括 app 和 Gradle Scripts 两个部分。其中，Gradle Scripts 中包含了一些应用程序在配置、编译和打包时产生的文件，这些文件在程序开发时基本用不到。因此，下面对 app 中一些常用的节点进行详细介绍。

1. manifests 节点

manifests 节点用于显示 Android 应用程序的配置。通常情况下，每个 Android 应用程序都必须包含 AndroidManifest.xml 文件，它位于 manifests 节点下，是整个 Android 应用的全局描述文件。在该文件内，需要标明应用的名称、使用的图标、Activity 和 Service 等信息，否则程序不能正常启动。例如，"Hellow World"应用中的 AndroidManifest.xml 文件如下：

```xml
<?xml version="1.0" encoding="utf-8"?>
<manifest xmlns:android="http://schemas.android.com/apk/res/android"
    package="com.example.helloworld">
<Application
android:allowBackup="true"
android:icon="@mipmap/ic_launcher"
android:label="@string/APP_name"
android:roundIcon="@mipmap/ic_launcher_round"
android:supportsRtl="true"
android:theme="@style/APPTheme">
<activity android:name=".MainActivity">
<intent-filter>
<action android:name="android.intent.action.MAIN" />
<category android:name="android.intent.category.LAUNCHER" />
</intent-filter>
</activity>
</Application>
</manifest>
```

除上述配置外，AndroidManifest.xml 文件中的其他重要元素及说明见表 1.1。

表 1.1　AndroidMainfest.xml 文件重要元素及说明

元素	说明
manifest	根节点，描述 package 中的所有内容
xmlns:android	包含命名空间的声明，其属性为 http://schemas.android.com/apl/res/android，表示 Android 中的各种标准属性能否在该 xml 文件中使用，提供了大部分元素中的数据
package	声明程序包
application	包含 package 中 Application 级别控件声明的根节点，一个 manifest 中可以包含零个或一个该元素
android:icon	应用程序图标
android:lable	应用程序标签
android:theme	应用程序采用的主题。默认为@style/APPTheme
activity	与用户交互的主要工具，是用户打开一个应用的初始界面
intent-filter	配置 intent 过滤器
action	控件支持的 intent action
category	控件支持的 intent category，这里通常用来指定应用程序默认的 activity

2. java 节点

java 节点显示包含的 Android 程序的所有包及源文件（.java）。例如，"Hello World"项目的 java 节点展开效果如图 1.24 所示。

图 1.24　java 节点展开效果

默认生成的 MainActivity.java 文件的关键代码如下：

```
package com.example.helloworld;
import android.support.v7.APP.APPCompatActivity;
import android.os.Bundle;
public class MainActivity extends APPCompatActivity {
@Override
protected void onCreate(Bundle savedInstanceState) {
super.onCreate(savedInstanceState);
    setContentView(R.layout.activity_main);
    }
}
</Application>
</manifest>
```

从上面的代码可以看出，Android Studio 创建的 MainActivity 类默认继承自 APPCompatActivity 类，并且在该类中重写了 Activity 类中的 onCreate 方法，在 onCreate 方法中通过 setContentView（R.layout.*activity_main*）方法设置当前的 Activity 要显示的布局文件为 *activity_main.xml*。

3. res 节点

res 节点用来显示保存在 res 目录下的资源文件。在 res 目录中还包含一些子目录，下面将对这些子目录进行详细说明。

（1）Drawable 子目录。

Drawable 子目录通常用来保存图片资源。

（2）Layout 子目录。

Layout 子目录主要用来存储 Android 程序中的布局文件。在创建 Android 程序时，系统会在此目录下默认生成一个 activity_main.xml 布局文件。

activity_main.xml 布局文件的关键代码如下：

```
<?xml version="1.0" encoding="utf-8"?>
<android.support.constraint.ConstraintLayout xmlns:android="http://schemas.android.com/apk/res/android"
    xmlns:APP="http://schemas.android.com/apk/res-auto"
    xmlns:tools="http://schemas.android.com/tools"
    android:layout_width="match_parent"
    android:layout_height="match_parent"
    tools:context="com.example.helloworld.MainActivity">
```

```
<TextView
android:layout_width="wrap_content"
android:layout_height="wrap_content"
android:text="Hello World!"
APP:layout_constraintBottom_toBottomOf="parent"
APP:layout_constraintLeft_toLeftOf="parent"
APP:layout_constraintRight_toRightOf="parent"
APP:layout_constraintTop_toTopOf="parent" />

</android.support.constraint.ConstraintLayout>
```

activity_main.xml 布局文件中的重要元素及说明见表 1.2。

表 1.2 activity_main.xml 布局文件中的重要元素及其说明

元素	说明
ConstraintLayout	布局管理器
xmlns:android	包含命名空间的声明,其属性值为http://schemas.android.com/apk/res/android,表示Android系统中的各种标准属性能够在该xml文件中使用,它提供了大部分元素中的数据,该属性一定不能省略
xmlns:tools	指定布局的默认工具
android:layout_width	指定当前视图在屏幕上显示的宽度
android:layout_height	指定当前视图在屏幕上显示的高度
TextView	文本框组件,用来显示文本
android: text	文本框组件显示的文本

另外,Android Studio 提供了可视化的布局编辑器来辅助用户开发布局文件。布局编辑器如图 1.25 所示。在该编辑器内,可以通过拖动组件来实现布局。

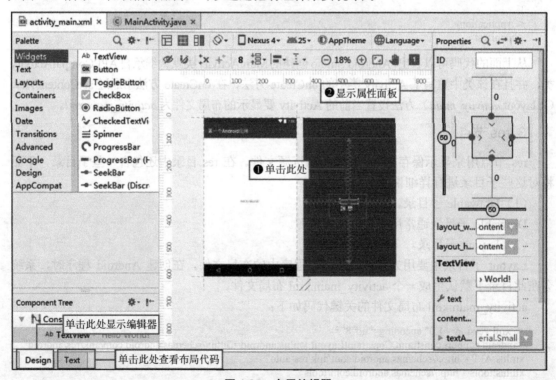

图 1.25 布局编辑器

(3) mipmap 子目录。

mipmap 子目录用于保存项目中应用程序的启动图标。为了保证良好的用户体验,需要为不同的分辨率提供不同的图片，并且分别存放在不同的目录中。通常情况下，Android Studio 会自动创建 mipmap-xxxhdpi（超超超高）、mipmap-xxhdpi（超超高）、mipmap-hdpe（超高）、mipmap-hdpi（高）和 mipmap-mdpi（中）五个目录，分别用于存放超超超高分辨率图片、超超高分辨率图片、超高分辨率图片、高分辨率图片和中分辨率图片，并且会自动创建对应五种分辨率的启动图标文件（ic_launcher.png），如图 1.26 所示。

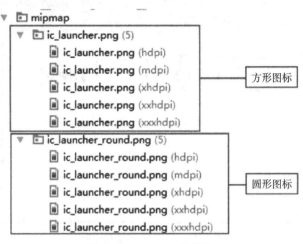

图 1.26　mipmap 子目录

（4）values 子目录。

values 子目录通常用于保存应用中使用的字符串、样式和颜色资源。例如，"Hello World" 的 values 子目录的结构如图 1.27 所示。

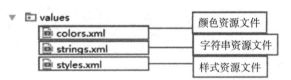

图 1.27　values 子目录的结构

各种资源保存在相应的 xml 文件中。例如，保存字符串的 Strings.xml 文件的代码如下：

```
01    <resources>
02        <string name="app_name">第一个 Android 应用</string>
03    </resources>
```

1.3.3　创建 Android 模拟器

Android 模拟器是 Google 公司官方提供的一款运行 Android 程序的虚拟机，可以模拟手机、平板电脑等设备。作为 Android 发人员，无论是否拥有基于 Android 系统的设备，都需要在 Android 模拟器上测试自己开发的 Android 程序。

由于启动 Android 模拟器需要配置 AVD，所以在运行 Android 程序之前，需要首先创建 AVD。创建 AVD 并启动 Android 模拟器的步骤如下。

　（说明：AVD 是 Android Virtual Device 的简称。通过 AVD 可以对 Android 模拟器进

行自定义配置,包括配置 Android 模拟器的硬件列表、模拟器的外观、支持的 Android 系统版本、附加 SDK 库和存储设置等。AVD 配置完成后,开发人员就可以按照这些配置来模拟真实的设备。)

(1)单击 Android Studio 工具栏上的 图标,弹出 AVD 管理器对话框,如图 1.28 所示。

图 1.28　AVD 管理器对话框

(2)单击"Create Virtual Device"按钮,弹出"Select Hardware"对话框,在该对话框中,选择要模拟的设备。例如,如果模拟 3.2 寸 HVGA 的设备,那么可以选择"32"HVGA slider(ADPI)",如图 1.29 所示。

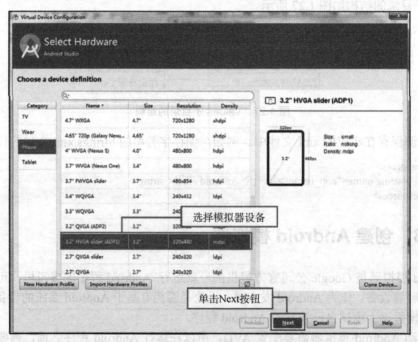

图 1.29　创建 AVD

(3)单击"Next"按钮,弹出选择系统镜像对话框,在该对话框中,默认情况下只能使用已下载的 ABI 为 x86 的系统镜像,并按照默认设置安装英特尔硬件加速器,如图 1.30 所示。

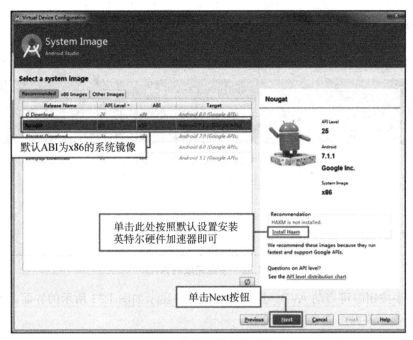

图 1.30　选择系统镜像对话框

（4）单击"Next"按钮，弹出验证配置对话框，在该对话框的"AVD Name"文本框中输入 AVD 名称，其他采用默认设置，如图 1.31 所示。

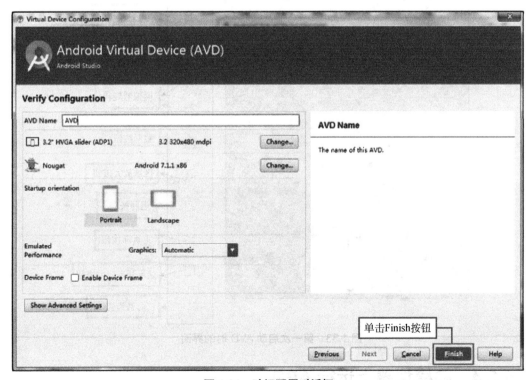

图 1.31　验证配置对话框

（5）单击"Finish"按钮完成 AVD 的创建。创建完成的 AVD 将显示在"Android Virtual

Device Manager"窗口中,如图 1.32 所示。

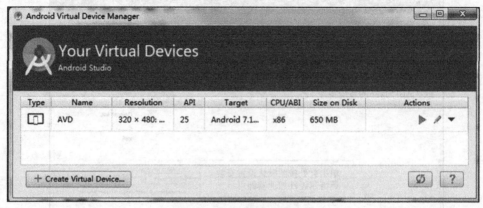

图 1.32 创建完成的 AVD

(6) 单击 ▶ 按钮即可启动 AVD。第一次启动时将显示如图 1.33 所示的界面。不同环境和不同配置,其显示界面可能会有所不同。

图 1.33 第一次启动 AVD 时的界面

1.3.4 在模拟器上运行 APP

创建 Android 应用程序后,还需要运行以查看其显示结果。使用模拟器来运行 Android

应用程序适用于不使用 Android 手机，或者测试各系统版本应用的兼容性的情况。在 Android Studio 中，通过模拟器运行"Hello World"的具体步骤如下。

（1）启动模拟器。

（2）在 Android Studio 主页工具栏中找到 app 下拉列表框，选择要运行的应用（这里为 APP），再单击右侧的 ▶ 按钮，弹出如图 1.34 所示的选择设备对话框。

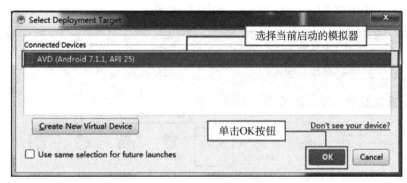

图 1.34 选择设备对话框

（3）选择当前启动的模拟器，单击"OK"按钮后，启动模拟器并运行程序。启动完毕后，在模拟器中将显示新创建的应用，运行效果如图 1.35 所示。

图 1.35 应用程序的运行效果

1.3.5 连接手机运行 APP

在上节中已经介绍了如何通过模拟器来运行 Android 应用程序，本节学习如何连接手机运行 Android 应用程序。该方法更符合用户场景，易于发现一些与硬件有关的问题，但是手机的系统很难与最新的 SDK 版本统一，所以对于 SDK 的一些新特性还是需要使用模拟器测试。

通过手机运行本书 1.3.4 节中编写的"Hello World",具体步骤如下。

(1) 将 Android 系统的手机连接到计算机上,通常情况下计算机桌面的右下角将提示正在安装设备驱动程序软件。

(2) 下载并安装"电脑管家"(或者"360 安全卫士",本书采用"电脑管家"),安装完成后"电脑管家"将自动显示主界面。在主界面中首先单击"工具箱"按钮,然后单击"应用宝"按钮进行"应用宝"的下载与安装,如图 1.36 所示。

图 1.36　安装"应用宝"

(3) "应用宝"安装完成后将自动显示如图 1.37 所示的授权计算机管理手机的对话框。

图 1.37　授权计算机管理手机对话框

同时,在手机上将显示如图 1.38 所示的是否允许 USB 调试的对话框。对于不同的手机,显示的允许 USB 调试的方式可能不同,只要根据自己的手机提示进行选择即可。

（4）在如图 1.38 所示的对话框中点击"确定"按钮，允许 USB 调试，在计算机中将弹出如图 1.39 所示计算机与手机进行连接的对话框。当出现如图 1.40 所示的连接成功窗口时则表示手机连接成功。

图 1.38　是否允许 USB 调试对话框

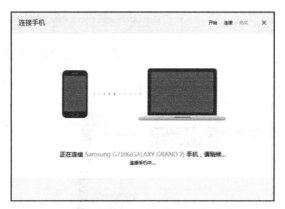

图 1.39　计算机与手机进行连接对话框

图 1.40　连接成功窗口

（5）返回 Android Studio 中，在工具栏中找到 app 下拉列表框，选择要运行的项目，再单击右侧的 按钮，将弹出如图 1.41 所示的选择要运行的设备对话框。选择相应设备，单击"OK"按钮运行项目。

图 1.41　选择要运行的设备对话框

（6）项目运行后，在手机上将显示应用的运行效果，如图 1.42 所示。

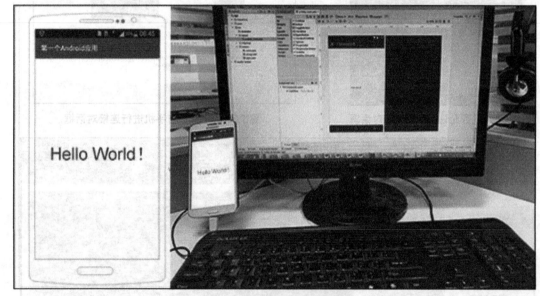

图 1.42　应用程序的运行效果

 ## 1.4　认识项目——购物商城 APP

1.4.1　开发背景

现在有很多人已经习惯了在工作或生活的间隙，把更多的"碎片"时间放在不断享受移动购物的乐趣上。为此，越来越多的企业都已经开始打造专属于自己企业的电子商务 APP。相信每个人的手机中都会有一个或几个电子商务 APP，如手机京东和手机淘宝等。本章将介绍的购物商城 APP 就是一款模仿手机京东实现的购物商城 APP。

1.4.2 系统功能设计

购物商城 APP 主要分为商城首页、商品分类、购物车和个人中心四大部分。其详细的功能结构如图 1.43 所示。

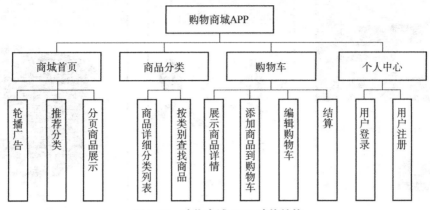

图 1.43 购物商城 APP 功能结构

1.4.3 项目包结构说明

在编写代码之前，可以首先把项目中可能用到的包（包结构）创建出来（例如，创建 activity 包，用于保存 APP 中所使用的 Activity），这样不仅可以方便以后的开发工作，而且还可以规范项目的整体架构。购物商城 APP 的项目包结构如图 1.44 所示。

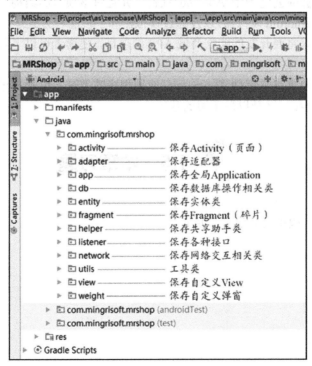

图 1.44 购物商城 APP 的项目包结构

1.4.4 系统预览

购物商城 APP 系统预览如图 1.45 至图 1.47 所示。

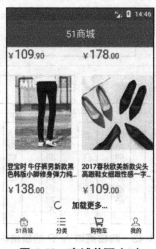
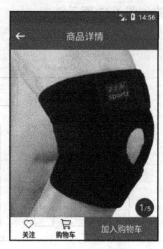

图 1.45　商城首页（1）　　　图 1.46　商城首页（2）　　　图 1.47　商品详情

1.5　本章小结

本章首先介绍了如何创建一个 Android 项目，并且对该项目的结构进行了详细的讲解。然后介绍了如何使用 Android 模拟器，以及通过模拟器和手机两种方式运行 Android 应用程序。最后介绍了本书接下来将要开发的购物商城 APP 的项目，使读者可以了解大体的开发框架和方向。

在阅读本章时，要重点掌握创建 Android 应用程序的方法，以及通过手机或模拟器运行 Android 应用程序的方法，这是以后学习的基础。

1.6　本章习题

如何创建 Android 模拟器并运行项目？

第 2 章　Android 用户界面设计

通过前面的学习，相信读者已经对 Android 系统有了一定的了解，本章将学习 Android 系统开发中一项很重要的内容——用户界面设计。Android 系统提供了多种控制 UI 界面的方法和布局方式，以及大量功能丰富的 UI 组件。通过这些组件，开发人员可以像搭积木一样，开发出优秀的用户界面。

本章知识框架如图 2.1 所示。

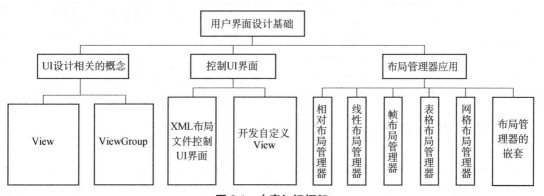

图 2.1　本章知识框架

 ## 2.1　UI 设计的相关概念

如果要开发的 Android 应用程序是运行在手机或平板电脑上的程序，那么这些程序给用户的第一印象就是用户界面，也就是 User Interface，简称 UI。用户界面是应用程序和用户之间进行信息交换的媒介。在 Android 系统中，设计用户界面又称 UI 设计，在进行 UI 设计时，经常会用到 View 和 ViewGroup。对于初识 Android 系统的人来说一般不好理解。下面将对这两个概念进行详细介绍。

2.1.1　View

View 在 Android 系统中可以理解为视图，它是占据在屏幕上的一个矩形区域，负责提供组件绘制和事件处理的方法。如果把 Android 应用程序的页面比喻为窗户，那么窗户中的每一块玻璃都相当于一个 View，如图 2.2 所示。View 类是所有的 UI 组件的基类。

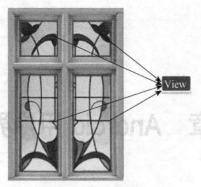

图 2.2　View 示意图

在 Android 系统中，View 类及其子类的相关属性既可以在 XML 布局文件中进行设置，也可以通过成员方法在 Java 代码中动态地设置。View 类支持的常用 XML 属性及对应的方法如表 2.1 所示。

表 2.1　View 类支持的常用 XML 属性及对应的方法

XML 属性	方　　法	描　　述
android:background	setBackGroundResource(int)	设置背景，其属性值为 Drawable 资源或颜色值
android:clickable	setClickable(boolean)	设置是否响应单击事件，其属性值为 Boolean 型的 True 或 False
android:elevation	setElevation（float）	这是 Android API21 新添加的，用于设置 Z 轴深度，其属性值为带单位的有效浮点数

2.1.2　ViewGroup

ViewGroup 在 Android 系统中代表容器。如果继续用窗户来比喻，那么 ViewGroup 就相当于窗户框，用于控制玻璃的安放，如图 2.3 所示。ViewGroup 类继承自 View 类，它是 View 类的拓展，是用来容纳其他组件的容器，但是由于 ViewGroup 是一个抽象类，所以在实际应用中通常作为使用 ViewGroup 的子类的容器。例如，在 2.3 节中将要介绍的布局管理器。

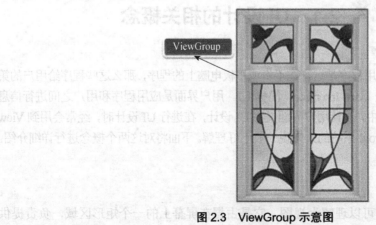

图 2.3　ViewGroup 示意图

ViewGroup 控制子组件的分布时（如设置子组件的内边距、宽度和高度等），还经常依赖

ViewGroup.LayoutParams 和 ViewGroup.MarginLayoutParams 两个内部类，下面分别进行介绍。

（1）ViewGroup. LayoutParams 类。

ViewGroup.LayoutParams 类封装了布局的位置、高度和宽度等信息。它支持 android: layout-height 和 android: layout-width 两个 XML 属性，它们的属性值可以使用精确的数值指定，也可以使用 FILL-PARENT（表示与父容器相同）、Match-parent（表示与父容器相同，需要 API 8 或以上版本支持）或 Warp-content（表示包裹其自身的内容）指定。

（2）ViewGroup.MarginLayoutParams 类。

ViewGroup.MarginLayoutParams 类用于控制其子组件的外边距。它支持的常用 XML 属性如表 2.2 所示。

表 2.2　ViewGroup. MarginLayoutParams 类支持的常用 XML 属性

XML 属性	描　　述
android:layout_marginBotton	设置底外边距
android:layout_marginEnd	设置结束边距
android_marginLeft	设置左外边距
android:layout_marginRight	设置右外边距
android:layout_marginStart	设置开始边距
android:layout_marginTop	设置顶外边距

在 Android 系统中，所有的 UI 界面都是由 View、ViewGroup 类及其子类组合而成的。在 ViewGroup 类中，除可以包含普通的 View 类外，还可以再次包含 ViewGroup 类。实际上，这使用了 Composite（组合）设计模式。View 类和 ViewGroup 类的层次结构如图 2.4 所示。

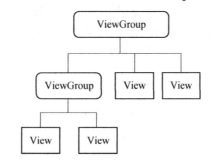

图 2.4　View 类和 ViewGroup 类的层次结构

 ## 2.2　控制 UI 界面

用户界面设计是 Android 应用程序开发的一项重要内容。在进行用户界面设计时，首先需要了解用何种方法控制 UI 界面。

2.2.1　使用 XML 布局文件控制 UI 界面

Android 系统提供了一种非常简单、方便的方法用于控制 UI 界面，该方法采用 XML 文件来进行界面布局，从而将布局界面的代码和逻辑控制的 Java 代码分离开来，使程序的结构更加清晰、明了。

使用 XML 布局文件控制 UI 界面可以分为以下两个关键步骤。

（1）在 Android 应用程序的 res\layout 目录下创建 XML 布局文件，该布局文件可以采用任何符合 Java 命名规则的文件名。

（2）在 Activity 中使用以下 Java 代码显示 XML 文件中布局的内容。

SetContentview（R.layout.activity_main）;

在上面的代码中，activity_main 是 XML 布局文件的文件名。

2.2.2 开发自定义的 View 类

一般情况下，开发 Android 应用程序的 UI 界面时，都不直接使用 View 类和 ViewGroup 类，而是使用这两个类的子类。例如，要显示一个图片，就可以使用 View 类的子类 ImageView。虽然 Android 系统提供了很多继承自 View 类的 UI 组件，但是在实际开发时还会出现不满足需要的情况。这时就可以通过继承 View 类来开发自己的组件。开发自定义的 View 组件大致分为以下三个步骤。

（1）创建一个继承自 android.view.View 类的 Java 类，并且重写构造方法。

（注意：在自定义的 View 类中，至少需要编写一个构造方法。）

（2）根据需要重写其他的方法。被重写的方法可以通过下面的方法找到。

在代码中单击鼠标右键，在弹出的快捷菜单中选择"Generate"菜单项，弹出如图 2.5 的快捷菜单，在该菜单中选择"Override Methods"菜单项，弹出如图 2.6 所示的选择进行覆盖或实现的方法对话框，在该对话框的列表中显示出可以被重写的方法。选择相应的方法，并单击"OK"按钮，Android Studio 将自动重写指定的方法。通常情况下，不需要重写全部的方法。

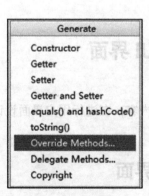

图 2.5 快捷菜单

图 2.6 选择进行覆盖或实现的方法对话框

（3）在项目的活动中，创建并实例化自定义 View 类，然后将其添加到布局管理器中。

2.3 布局管理器

在 Android 系统中，每一个组件在窗体中都有具体的位置和大小，因此在窗体中摆放各种组件时很难进行判断。不过，使用 Android 布局管理器可以很方便地控制各组件的位置和大小。Android 系统提供了以下五种布局管理器。

◆ 相对布局管理器（RelativeLayout）：通过相对定位的方式来控制组件的摆放位置。
◆ 线性布局管理器（LinearLayout）：在垂直或水平方向上依次摆放组件。
◆ 帧布局管理器（FrameLayout）：没有任何定位方式，默认情况下，所有的组件都会摆放在容器的左上角，逐个覆盖。
◆ 表格布局管理器（TableLayout）：使用表格的方式按行和列来摆放组件。
◆ 绝对布局管理器（AbsoluteLayout）：通过绝对坐标（x、y）的方式控制组件的摆放位置。

其中，绝对布局管理器在 Android 2.0 版本中被标记为已过期，但是可以使用帧布局管理器或相对布局管理器替代。另外，在 Android 4.0 版本以后，又提供了一个新的布局管理器——网格布局管理器（GridLayout），通过它可以实现跨行或跨列摆放组件。

Android 系统提供的布局管理器均直接或间接地继承自 ViewGroup 类，如图 2.7 所示。因此，所有的布局管理器都可以作为容器使用，可以向管理器中添加 UI 组件。当然，也可以将一个或多个布局管理器嵌套到其他的布局管理器中，在本章的 2.3.6 小节将介绍布局管理器的嵌套。

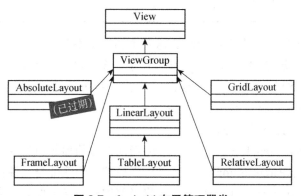

图 2.7 Android 布局管理器类

2.3.1 相对布局管理器

相对布局管理器是通过相对定位的方式来控制组件出现在布局的任何位置。例如，如图 2.8 所示的页面就是采用相对布局管理器来进行布局的，在界面中首先放置组件 A，然后放置组件 B，使其位于组件 A 的下方，再放置组件 C，使其位于组件 A 的下方且位于组件 B 的右侧。

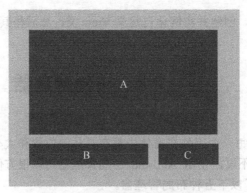

图 2.8 相对布局管理器布局示意

在 Android 系统中可以在 XML 布局文件中定义相对布局管理器,也可以使用 Java 代码来创建。本书推荐在 XML 布局文件中定义相对布局管理器。在 XML 布局文件中,定义相对布局管理器可以使用<RelativeLayout>标记,其基本的语法格式如下:

```
<RelativeLayout xmlns:android="http://schemas.android.com/apk/res/android"
    属性列表
>
</Relativelayout>
```

在上面的语法中,<RelativeLayout>为起始标记,</RelativeLayout>为结束标记。在起始标记中的 xmlns:android 为设置 XML 命名空间的属性,其属性值为固定写法。

RelativeLayout 常用 XML 属性如表 2.3 所示。

表 2.3 RelativeLayout 常用 XML 属性

XML 属性	描述
android:gravity	用于设置布局管理器中各子组件的对齐方式
android:ignoreGravity	用于指定哪个组件不受 gravity 属性的影响

在相对布局管理器中,只有上面介绍的这两个属性是不够的,为了更好地控制该布局管理器中各子组件的布局分布,RelativeLayout 提供了一个内部类 RelativeLayout.LayoutParams,通过该类提供的大量 XML 属性,可以很好地控制相对布局管理器中各组件的分布方式。RelativeLayout. LayoutParams 支持的 XML 属性分类如下所示。

第一类:属性值为 True 或 False。

android:layout_alignParentTop   将当前控件的顶部与父控件的顶部对齐
android:layout_alignParentBottom   将当前控件的底部与父控件的底部对齐
android:layout_alignParentEnd   将当前控件与父控件的结束位置对齐
android:layout_alignParentStart   将当前控件与父控件的起始位置对齐
android:layout_alignParentLeft   将当前控件的左边缘与父控件的左边缘对齐
android:layout_alignParentRight   将当前控件的右边缘与父控件的右边缘对齐
android:layout_centerInParent   将当前控件放在父控件的中心(水平居中+垂直居中)
android:layout_centerVertical   在父控件中垂直居中
android:layout_centerHorizontal   在父控件中水平居中

第二类：属性值必须为 ID 的引用名"@id/id-name"，即给定控件。

属性	说明
android:layout_alignBaseLine	将当前控件与给定控件的基准线对齐
android:layout_alignTop	将当前控件的顶部与给定控件的顶部对齐
android:layout_alignBottom	将当前控件的底部与给定控件的底部对齐
android:layout_alignLeft	将当前控件的左边缘与给定控件的左边缘对齐
android:layout_alignRight	将当前控件的右边缘与给定控件的右边缘对齐
android:layout_above	将当前控件放在给定控件的上方
android:layout_below	将当前控件放在给定控件的下方
android:layout_toLeftOf	将当前控件放在给定控件的左侧
android:layout_toRightOf	将当前控件放在给定控件的右侧

第三类：属性值为具体的像素值，如 30dip、40px 等。

属性	说明
android:layout_margin	指定当前控件的外边距
android:layout_marginBottom	指定当前控件底部与外部控件之间的距离
android:layout_marginLeft	指定当前控件左边缘与外部控件之间的距离
android:layout_marginRight	指定当前控件右边缘与外部控件之间的距离
android:layout_marginTop	指定当前控件顶部与外部控件之间的距离

2.3.2 线性布局管理器

线性布局管理器用于将放入其中的组件按照垂直或水平方向来布局，也就是控制放入其中的组件横向排列或纵向排列。其中，纵向排列的称为垂直线性布局管理器，如图 2.9 所示；横向排列的称为水平线性布局管理器，如图 2.10 所示。在垂直线性布局管理器中，每一行只能放一个组件，而在水平线性布局管理器中每一列中只能放一个组件。另外，Android 系统中的线性布局管理器中的组件不会换行，当组件一个挨着一个排列到窗体的边缘后，剩下的组件将不会被显示出来。

图 2.9 垂直线性布局管理器

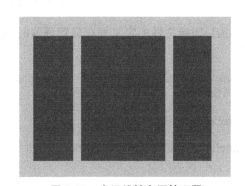
图 2.10 水平线性布局管理器

在 XML 布局文件中定义线性布局管理器时需要使用<LinearLayout>标记，其基本的语法格式如下：

```
<LinearLayout xmlns:android="http://schemas.android.com/apk/res/android"
    属性列表
>
</LinearLayout>
```

LinearLayout 常用 XML 属性如表 2.4 所示。

表 2.4 LinearLayout 常用 XML 属性

XML 属性	描述
android:orientation	设置线性布局的方向，将值设置为 horizontal 表示水平排列，设置为 vertical 表示垂直排列，默认为 horizontal
android:layout_weight	指定当前控件在 LinearLayout 中所占的权重

2.3.3 帧布局管理器

在帧布局管理器中，每加入一个组件都将创建一个空白的区域，通常称为一帧，默认情况下，这些帧都会被放在屏幕的左上角，即帧布局从屏幕的左上角坐标点（0,0）开始布局。如果多个组件层叠排序，则后面的组件覆盖将前面的组件，如图 2.11 所示。

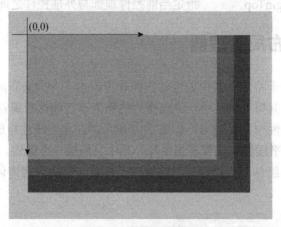

图 2.11 帧布局管理器

在 XML 布局文件中定义帧布局管理器时需要使用<FrameLayout>标记，其基本的语法格式如下：

```
<FrameLayout xmlns:android="http://schemas.android.com/apk/res/android"
    属性列表
>
</ FrameLayout >
```

FrameLayout 常用 XML 属性如表 2.5 所示。

表 2.5　FrameLayout 常用 XML 属性

XML 属性	描　　述
android:layout_gravity	设置在父控件中显示的位置
android:foreground	设置前景图像
android:foregroundGravity	设置前景图像显示的位置

2.3.4　表格布局管理器

表格布局管理器与常见的表类似，它以行、列的形式来管理放入其中的 UI 组件，如图 2.12 所示。表格布局管理器使用<TableLayout>标记定义，在表格布局管理器中，可以添加多个<TableRow>标记，每个<TableRow>标记占用一行。由于<TableRow>标记也是容器，所以在该标记中还可添加其他组件。在<TableRow>标记中，每添加一个组件，表格就会增加一列。在表格布局管理器中，列既可以被隐藏，也可以被设置为伸展的，从而填充可利用的屏幕空间，还可以设置为强制收缩，直到表格匹配屏幕大小。

图 2.12　表格布局管理器

在 XML 布局文件中定义表格布局管理器的基本语法格式如下：

```
< TableLayout xmlns:android="http://schemas.android.com/apk/res/android"
属性列表
    >
< TableRow  属性列表>需要添加的 UI 组件</TableRow >
多个< TableRow >
</ TableLayout >
```

TableLayout 继承自 LinearLayout，因此它完全支持 LinearLayout 所支持的全部 XML 属性，此外，TableLayout 还支持如表 2.6 所示的 XML 属性。

表 2.6　TableLayout 常用 XML 属性

XML 属性	描　　述
android:collapseColumns	设置要隐藏的列的列序号（序号从 0 开始），多个列序号用逗号（,）分隔
android:shrinkColumns	设置可收缩的列的列序号（序号从 0 开始），多个列序号用逗号（,）分隔
android:stretchColumns	设置可伸展的列序号（序号从 0 开始），多个列序号用逗号（,）分隔

2.3.5 网格布局管理器

网格布局管理器是在 Android 4.0 版本中提出的,用 GridLayout 表示。在网格布局管理器中,屏幕被虚拟的细线划分成行、列和单元格,每个单元格放置一个组件,该组件也可以跨行或跨列摆放,如图 2.13 所示。

图 2.13 网格布局管理器

在 XML 布局文件中定义网格布局管理器时可使用<GridLayout>标记,其基本的语法格式如下:

```
< GridLayout xmlns:android="http://schemas.android.com/apk/res/android"
属性列表
    >
</GridLayout>
```

GridLayout 常用的 XML 属性如表 2.7 所示。

表 2.7 GridLayout 常用 XML 属性

XML 属性	描 述
android:columnCount	设置网格最大列数
android:rowCount	设置网格最大行数
android:orientation	GridLayout 中子元素的布局方向
android:alignmentMode	alignBounds:按子视图内容对齐;alignMargins:按子视图边距对齐,此为默认值
android:columnOrderPreserved	使列边界显示的顺序和列索引的顺序相同,默认值是 True
android:rowOrderPreserved	使行边界显示的顺序和行索引的顺序相同,默认值是 True
android:useDefaultMargins	没有指定视图的布局参数时使用默认的边距,默认值是 False

2.3.6 布局管理器的嵌套

在进行用户界面设计时,很多时候只通过一种布局管理器很难实现想要的界面效果,这时就需要混合使用多种布局管理器，即布局管理器的嵌套。在实现布局管理器的嵌套时，只需记住以下几点原则即可。

（1）根布局管理器必须包含 xmlns 属性。

（2）在一个布局文件中，最多只能有一个根布局管理器。如果想要使用多个布局管理器，就需要使用一个根布局管理器将它们括起来。

（3）不能嵌套太深。如果嵌套太深，则会影响性能，并会降低页面的加载速度。

2.4 购物商城 APP 的布局设计

2.4.1 购物商城 APP 商城首页布局

访问购物商城 APP 时，首先进入的就是购物商城 APP 主框架页面。主框架页面主要分为如图 2.14 所示的两部分，一部分是切换界面部分，另一部分是底部选择栏。其中，切换界面部分通过帧布局管理器（FrameLayout）来实现；底部选择栏采用包含另一个布局文件 bottombar_layout.xml 的方式来实现。

图 2.14　商城首页布局

在开始之前，首先把准备好的图标等素材放入 res\drawable 目录下，如图 2.15 所示，素材说明见本书附录。

图 2.15 素材目录

activity_main.xml 具体代码如下：

```xml
<?xml version="1.0" encoding="utf-8"?>
<LinearLayout
    xmlns:android="http://schemas.android.com/apk/res/android"
    xmlns:tools="http://schemas.android.com/tools"
    android:id="@+id/activity_main"
    android:layout_width="match_parent"
    android:layout_height="match_parent"
    android:orientation="vertical"
    tools:context=".activity.MainActivity">
    <!-- 切换界面部分 -->
    <FrameLayout
        android:id="@+id/frag_home"
        android:layout_width="match_parent"
        android:layout_height="0dp"
        android:layout_weight="1"/>
    <!-- 底部选择栏 -->
    <include layout="@layout/bottombar_layout"/>
</LinearLayout>
```

创建一个新的布局 xml 文件，命名为 bottombar_layout.xml，具体代码如下：

```xml
<?xml version="1.0" encoding="utf-8"?>
<RelativeLayout
    xmlns:android="http://schemas.android.com/apk/res/android"
    xmlns:tools="http://schemas.android.com/tools"
    style="@style/MineBottomBarStyle"
    android:elevation="2dp"
    android:translationZ="2dp"
    tools:showIn="@layout/activity_main">
    <RadioGroup
        android:id="@+id/bottombar"
        style="@style/MineBottomBarStyle"
        tools:showIn="@layout/activity_main">
        <!-- 商城首页面 -->
        <RadioButton
            android:id="@+id/mr_shoppingmall"
```

```xml
            android:drawableTop="@drawable/mr_shoppingmall"
            android:text="首页"
            style="@style/MineBottomBarButtonStyle" />
        <!-- 分类 -->
        <RadioButton
            android:id="@+id/mr_category"
            android:drawableTop="@drawable/mr_category"
            android:text="分类"
            style="@style/MineBottomBarButtonStyle" />
        <!-- 购物车 -->
        <RadioButton
            android:id="@+id/mr_shoppingcart"
            android:drawableTop="@drawable/mr_shoppingcart"
            android:text="购物车"
            style="@style/MineBottomBarButtonStyle" />
        <!-- 我的 -->
        <RadioButton
            android:id="@+id/mr_mine"
            android:drawableTop="@drawable/mr_mine"
            android:text="我的"
            style="@style/MineBottomBarButtonStyle" />
    </RadioGroup>
    <!-- 遮罩 -->
    <LinearLayout
        android:layout_width="match_parent"
        android:layout_height="match_parent">
        <TextView
            android:layout_width="0dp"
            android:layout_height="match_parent"
            android:layout_weight="1" />
        <TextView
            android:layout_width="0dp"
            android:layout_height="match_parent"
            android:layout_weight="1" />
        <TextView
            android:id="@+id/show_count"
            android:layout_width="0dp"
            android:layout_height="match_parent"
            android:layout_weight="1" />
        <TextView
            android:layout_width="0dp"
            android:layout_height="match_parent"
            android:layout_weight="1" />
    </LinearLayout>
</RelativeLayout>
```

定义一些样式，由于其他地方也会用到这些样式，所以将其编写在公用的 styles.xml 配置文件中，具体代码如下：

```xml
<resources>
    <!-- 界面的样式风格 -->
    <style name="APPTheme" parent="Theme.APPCompat.Light.NoActionBar">
        <item name="colorPrimary">@color/colorPrimary</item>
        <item name="colorPrimaryDark">@color/colorPrimaryDark</item><!--状态栏背景颜色-->
        <item name="colorAccent">@color/colorAccent</item>
    </style>
    <!-- 标题栏的样式 -->
```

```xml
<style name="MineTitleBarStyle">
    <item name="android:layout_height">?attr/actionBarSize</item>
    <item name="android:layout_width">match_parent</item>
    <item name="android:background">@color/title_bar_background</item>
    <item name="android:paddingTop">3dp</item>
    <item name="android:paddingBottom">3dp</item>
    <item name="android:paddingLeft">10dp</item>
    <item name="android:paddingRight">10dp</item>
</style>
<!-- 底部选择栏的样式 -->
<style name="MineBottomBarStyle">
    <item name="android:orientation">horizontal</item>
    <item name="android:layout_width">match_parent</item>
    <item name="android:gravity">center_vertical</item>
    <item name="android:layout_height">?attr/actionBarSize</item>
    <item name="android:paddingTop">2dp</item>
    <item name="android:paddingBottom">2dp</item>
    <item name="android:background">@color/white</item>
</style>
<!-- 底部选择栏按钮的样式 -->
<style name="MineBottomBarButtonStyle">
    <item name="android:layout_width">0dp</item>
    <item name="android:layout_height">wrap_content</item>
    <item name="android:layout_weight">1</item>
    <item name="android:button">@null</item>
    <item name="android:gravity">center</item>
    <item name="android:textSize">12dp</item>
    <item name="android:textColor">@drawable/bottombar_text_selector_color</item>
</style>
<!-- 一个固定边距 -->
<style name="MarginStyle">
    <item name="android:layout_marginLeft">10dp</item>
    <item name="android:layout_marginRight">10dp</item>
    <item name="android:layout_marginTop">5dp</item>
</style>
<!-- 一个固定边距 -->
<style name="PaddingStyle">
    <item name="android:paddingLeft">10dp</item>
    <item name="android:paddingRight">10dp</item>
    <item name="android:paddingTop">5dp</item>
</style>
<!-- 背景不模糊的弹窗效果 -->
<style name="Custom_Dialog_Theme_Background_DimEnabled_False" parent="android:style/Theme.Dialog">
    <!--背景颜色及和透明程度-->
    <item name="android:windowBackground">@android:color/transparent</item>
    <!--是否去除标题 -->
    <item name="android:windowNoTitle">true</item>
    <!--是否去除边框-->
    <item name="android:windowFrame">@null</item>
    <!--是否浮现在activity之上-->
    <item name="android:windowIsFloating">true</item>
    <!--是否模糊-->
    <item name="android:backgroundDimEnabled">false</item>
</style>
<!-- 背景模糊的弹窗效果 -->
<style name="Custom_Dialog_Theme_Background_DimEnabled_True" parent="android:style/Theme.Dialog">
```

```xml
        <!--背景颜色及和透明程度-->
        <item name="android:windowBackground">@android:color/transparent</item>
        <!--是否去除标题 -->
        <item name="android:windowNoTitle">true</item>
        <!--是否去除边框-->
        <item name="android:windowFrame">@null</item>
        <!--是否浮现在 activity 之上-->
        <item name="android:windowIsFloating">true</item>
        <!--是否模糊-->
        <item name="android:backgroundDimEnabled">true</item>
    </style>
    <!-- 弹窗后退出动画 -->
    <style name="Custom_Dialog_Anim_Style">
        <item name="android:windowEnterAnimation">@anim/custom_dialog_in</item>
        <item name="android:windowExitAnimation">@anim/custom_dialog_out</item>
    </style>
    <!-- 主页分类按钮的样式 -->
    <style name="Home_Type_Style">
        <item name="android:layout_width">0dp</item>
        <item name="android:layout_height">match_parent</item>
        <item name="android:layout_weight">1</item>
        <item name="android:textColor">@color/black</item>
        <item name="android:gravity">center_horizontal</item>
        <item name="android:button">@null</item>
    </style>

    <style name="Mine_Item_Style">
        <item name="android:layout_marginTop">10dp</item>
        <item name="android:layout_width">match_parent</item>
        <item name="android:layout_height">wrap_content</item>
        <item name="android:paddingLeft">10dp</item>
        <item name="android:paddingRight">10dp</item>
        <item name="android:paddingTop">5dp</item>
        <item name="android:paddingBottom">5dp</item>
        <item name="android:background">@color/white</item>
    </style>

    <style name="Mine_Image_Style">
        <item name="android:layout_width">wrap_content</item>
        <item name="android:layout_height">wrap_content</item>
        <item name="android:layout_marginRight">10dp</item>
        <item name="android:layout_gravity">center_vertical</item>
    </style>

    <style name="Mine_Text_Style">
        <item name="android:layout_width">0dp</item>
        <item name="android:layout_weight">1</item>
        <item name="android:layout_height">wrap_content</item>
        <item name="android:textColor">@color/black</item>
        <item name="android:textSize">16dp</item>
        <item name="android:layout_gravity">center_vertical</item>
    </style>
</resources>
```

运行效果如图 2.16 所示。

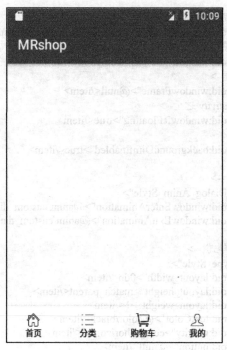

图 2.16 运行效果

新建一个布局并命名为 fragment_home.xml，用于展示首页内容。该页面最外层是线性布局，里面嵌套了两个相对布局，相对布局的背景图片为相应的广告图片。两个相对布局的中间是一个线性布局，其中水平排列四个 RadioButton 分类选择按钮，具体代码如下：

```xml
<?xml version="1.0" encoding="utf-8"?>
<LinearLayout xmlns:android="http://schemas.android.com/apk/res/android"
    android:orientation="vertical" android:layout_width="match_parent"
    xmlns:tools="http://schemas.android.com/tools"
    android:layout_height="match_parent">
    <RelativeLayout
        android:layout_width="match_parent"
        android:layout_height="@dimen/dp_160"
        android:background="@drawable/banner5">
    </RelativeLayout>
    <!-- 分类选择 -->
    <LinearLayout
        android:layout_width="match_parent"
        android:layout_height="wrap_content"
        android:layout_marginBottom="@dimen/dp_5"
        android:layout_marginTop="@dimen/dp_5">

        <RadioButton
            android:id="@+id/shop_type1"
            style="@style/Home_Type_Style"
            android:drawableTop="@drawable/mr_shop_type1"
            android:text="衣服裤子" />
        <RadioButton
            android:id="@+id/shop_type2"
            style="@style/Home_Type_Style"
            android:drawableTop="@drawable/mr_shop_type2"
            android:text="精品鞋子"
```

```xml
            <RadioButton
                android:id="@+id/shop_type3"
                style="@style/Home_Type_Style"
                android:drawableTop="@drawable/mr_shop_type3"
                android:text="食品饮料"/>
            <RadioButton
                android:id="@+id/shop_type4"
                style="@style/Home_Type_Style"
                android:drawableTop="@drawable/mr_shop_type4"
                android:text="美妆个护"/>
        </LinearLayout>
        <RelativeLayout
            android:layout_width="match_parent"
            android:layout_height="wrap_content"
            android:background="@drawable/banner6">
        </RelativeLayout>

</LinearLayout>
```

在后台 MainActivity.java 文件中修改 MainActivity 的 onCreate 方法，更改加载布局的 ID 为 fragment_home。

```java
@Override
protected void onCreate(Bundle savedInstanceState) {
    super.onCreate(savedInstanceState);
    //setContentView(R.layout.activity_main);
    setContentView(R.layout.fragment_home);

}
```

单击 AndroidStudio 右上方的 Run 按钮，选择模拟器运行项目，运行效果如图 2.17 所示。

图 2.17 商城首页

2.4.2 个人中心页面布局

每个用户软件都有一个用户的个人中心,以方便用户进行登录、注销及其他相关操作。上一节介绍了商城首页布局,接下来介绍用户的个人中心页面布局。个人中心页面布局如图 2.18 所示。

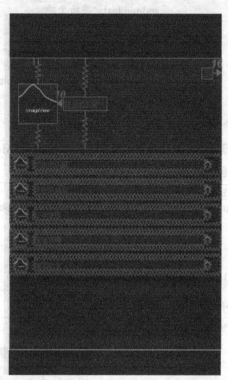

图 2.18 个人中心布局

新建 XML 布局并命名为 fragment_mine.xml。模仿京东商城的个人中心页面,最外层包裹一个线性布局,其内部分为上下两部分。

上部分是一个相对布局,里面放置一个 ImageView 组件(用于显示用户头像),一个 TextView 组件(用于显示用户的登录状态),以及一个线性布局用于放置右上角的设置按钮,具体代码如下:

```
<?xml version="1.0" encoding="utf-8"?>
<LinearLayout xmlns:android="http://schemas.android.com/apk/res/android"
    android:orientation="vertical" android:layout_width="match_parent"
    android:layout_height="match_parent">
    <RelativeLayout
        android:layout_width="match_parent"
        android:layout_height="@dimen/dp_160"
        android:background="@drawable/mr_mine_background">
        <!-- 设置头像 -->
        <ImageView
            android:id="@+id/custom_image"
            android:layout_width="wrap_content"
            android:layout_height="wrap_content"
```

```xml
            android:layout_centerVertical="true"
            android:src="@drawable/mr_header"
            android:layout_marginLeft="@dimen/sp_16"/>
        <!-- 登录注册 -->
        <TextView
            android:id="@+id/custom_login"
            android:layout_width="wrap_content"
            android:layout_height="wrap_content"
            android:layout_toRightOf="@id/custom_image"
            android:layout_centerVertical="true"
            android:text="登录/注册"
            android:textSize="@dimen/sp_18"
            android:textColor="@color/white"
            android:layout_marginLeft="@dimen/dp_10"/>
        <!-- 设置 -->
        <LinearLayout
            android:layout_alignParentRight="true"
            android:layout_width="wrap_content"
            android:layout_marginRight="@dimen/dp_16"
            android:layout_marginTop="@dimen/dp_16"
            android:layout_height="wrap_content">
            <!-- 设置按钮 -->
            <ImageButton
                android:id="@+id/custom_setting"
                android:layout_width="wrap_content"
                android:layout_height="wrap_content"
                android:src="@drawable/setting_selector"
                android:background="@null"/>
        </LinearLayout>
    </RelativeLayout>

</LinearLayout>
```

下部分则是一些线性布局，展示商城一些常用功能的入口，具体代码如下：

```xml
    <!-- 关注的店铺 -->
    <LinearLayout style="@style/Mine_Item_Style">
        <ImageView
            android:src="@drawable/mr_mine1"
            style="@style/Mine_Image_Style" />
        <TextView
            android:text="关注的店铺"
            style="@style/Mine_Text_Style" />
        <ImageView
            android:src="@drawable/mr_right_to"
            style="@style/Mine_Image_Style" />
    </LinearLayout>
    <!-- 关注的商品 -->
    <LinearLayout style="@style/Mine_Item_Style">
        <ImageView
            android:src="@drawable/mr_mine2"
            style="@style/Mine_Image_Style" />
        <TextView
            android:text="关注的商品"
            style="@style/Mine_Text_Style" />
        <ImageView
            android:src="@drawable/mr_right_to"
            style="@style/Mine_Image_Style" />
    </LinearLayout>
```

```xml
<!-- 订单详情 -->
<LinearLayout style="@style/Mine_Item_Style">
    <ImageView
        android:src="@drawable/mr_mine3"
        style="@style/Mine_Image_Style" />
    <TextView
        android:text="订单详情"
        style="@style/Mine_Text_Style" />
    <ImageView
        android:src="@drawable/mr_right_to"
        style="@style/Mine_Image_Style" />
</LinearLayout>
<!-- 快递查询 -->
<LinearLayout style="@style/Mine_Item_Style">
    <ImageView
        android:src="@drawable/mr_mine4"
        style="@style/Mine_Image_Style" />
    <TextView
        android:text="快递查询"
        style="@style/Mine_Text_Style" />
    <ImageView
        android:src="@drawable/mr_right_to"
        style="@style/Mine_Image_Style" />
</LinearLayout>
<!-- 浏览记录 -->
<LinearLayout style="@style/Mine_Item_Style">
    <ImageView
        android:src="@drawable/mr_mine5"
        style="@style/Mine_Image_Style" />
    <TextView
        android:text="浏览记录"
        style="@style/Mine_Text_Style" />
    <ImageView
        android:src="@drawable/mr_right_to"
        style="@style/Mine_Image_Style" />
</LinearLayout>
```

修改 MainActivity 的 onCreate 方法，更改加载布局的 ID 为 fragment_mine，具体代码如下：

```java
@Override
protected void onCreate(Bundle savedInstanceState) {
    super.onCreate(savedInstanceState);
    //主要框架页面
    //setContentView(R.layout.activity_main);
    //首页页面
    //setContentView(R.layout.fragment_home);
    //个人中心页面
    setContentView(R.layout.fragment_mine);
}
```

保存后，单击 AndroidStudio 右上方的 Run 按钮，选择模拟器运行项目，运行效果如图 2.19 所示。

图 2.19 个人中心页面运行效果

 2.5 本章小结

本章首先介绍了什么是 UI 界面，以及与 UI 设计相关的概念；然后介绍了控制 UI 界面的两种方法，一种是使用 XML 布局文件控制，另一种是开发自定义的 View 类；接下来又介绍了五种常用的布局管理器的基本使用方法，以及如何进行布局管理器的嵌套。在五种常用的布局管理器中，相对布局管理器和线性布局管理器最为常用，需要重点掌握。

 2.6 本章习题

网格布局管理器和表格布局管理的区别是什么？

第 3 章　常用 UI 组件

组件是 Android 系统应用程序设计的基本组成单位，通过使用组件可以高效地开发 Android 应用程序。熟练掌握组件的使用是合理、有效地进行 Android 应用程序开发的重要前提。本章将对 Android 系统中提供的常用组件进行详细介绍。本章知识框架如图 3.1 所示。

图 3.1　本章知识框架

3.1　常用组件

3.1.1　文本类组件

Android 系统提供了一些与文本的显示和输入相关的组件，通过这些组件可以显示或输入文本。其中，用于显示文本的组件称为文本框组件，用 TextView 类表示；用于编辑文本的组件称为编辑框组件，用 EditText 类表示。这两个组件的最大区别是 TextView 不允许用户编辑文本内容，而 EditText 则允许用户编辑文本内容。它们之间的继承关系如图 3.2 所示。

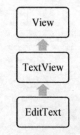

图 3.2　文本类组件继承关系

从图 3.2 中可以看出，TextView 组件继承自 View，而 EditText 组件又继承自 TextView 组件。

3.1.1.1　TextView 组件

Android 系统官网推荐在 XML 布局文件中定义文本框组件，TextView 的 XML 布局方式如图 3.3 所示。

```
<TextView
    android:layout_width="fill_parent"
    android:layout_height="wrap_content"
    android:textSize="16sp"-------------------字体大小
    android:textColor="#ffffff"
    android:padding="10dip"--------------------组件周围空隙大小
    android:background="#cc0000"-----------背景颜色
    android:layout_gravity="center"
    android:text ="Android Studio 工具箱"/>
```

图 3.3　TextView 的 XML 布局方式

TextView 常用的 XML 属性如表 3.1 所示。

表 3.1　TextView 常用的 XML 属性

属　　性	描　　述
android:id	这是唯一标识控件的 ID
android:capitalize	指定该 TextView 中有一个文本输入法会自动利用什么类型的用户 0：不要自动大写任何文本； 1：大写每句的第一个字； 2：大写每个单词的第一个字母； 3：大写每个字符
android:cursorVisible	使光标可见（默认）或不可见，默认值为 False
android:editable	如果设置为 True，则指定 TextView 的一个输入法
android:fontFamily	字体系列（由字符串命名）的文本
android:gravity	指定排列由视图的 x 和/或 y 轴的文本时，该文本比视图小
android:hint	提示文本显示文本为空
android:inputType	数据的类型被放置在一个文本字段中，如手机、日期、时间、号码、密码等
android:maxHeight	使 TextView 至多到多少像素高
android:maxWidth	使 TextView 至多到多少像素宽
android:minHeight	使 TextView 至少到多少像素高
android:minWidth	使 TextView 至少到多少像素宽
android:password	是否使用掩码显示输入的密码，值为 True 或 False
android:phoneNumber	如果设置，则指定 TextView 具有一个电话号码的输入法，可能的值是 True 或 False
android:text	要显示的文字
android:textAllCaps	目前所有大写的文本，可能的值是 True 或 False
android:textColor	文本颜色，可以是一个颜色值，形式为"#rgb"、"#argb"、"#rrggbb"或"#aarrggbb"

续表

属　性	描　述
android:textColorHint	颜色的提示文字，可以是一个颜色值，形式为"#rgb"、"#argb"、"#rrggbb"或"#aarrggbb"
android:textIsSelectable	表示可被选择的非可编辑的文本的内容，可能的值是 True 或 False
android:textSize	大小的文字。文字推荐尺寸类型是"sp"的比例像素（如15sp）
android:textStyle	样式（粗体、斜体、BOLDITALIC）的文本。可以使用"|"分隔值。 0：normal； 1：bold； 2：italic
android:typeface	字体（正常，SANS，衬线字体，等宽）的文本。可以使用"|"分隔值。 0：normal； 1：sans； 2：serif； 3：monospace

3.1.1.2　EditText 组件

在 Android 应用程序中，编辑框组件的应用非常普遍。例如，如图 3.4 所示的登录页面布局。布局中"请输入账号"的编辑框组件代码如下：

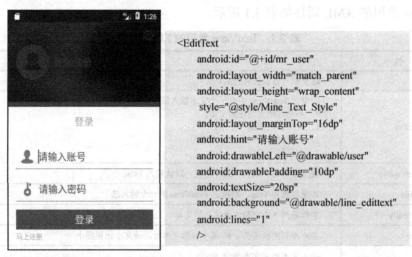

```
<EditText
    android:id="@+id/mr_user"
    android:layout_width="match_parent"
    android:layout_height="wrap_content"
    style="@style/Mine_Text_Style"
    android:layout_marginTop="16dp"
    android:hint="请输入账号"
    android:drawableLeft="@drawable/user"
    android:drawablePadding="10dp"
    android:textSize="20sp"
    android:background="@drawable/line_edittext"
    android:lines="1"
    />
```

图 3.4　登录页面布局

EditText 基本语法格式如下：

```
<EditText
属性列表
/>
```

3.1.2　按钮类组件

Android 系统提供了一些按钮类的组件，主要包括普通按钮、图片按钮、单选按钮和复选框按钮等。其中，普通按钮使用 Button 类表示，用于触发一个指定的事件；图片按钮使用 ImageButton 表示，也用于触发一个指定的事件，只不过该按钮以图片的形式表现；单选按钮使用 RadioButton 类表示；复选框按钮使用 CheckBox 类表示。按钮类组件的继承关系如图 3.5 所示。

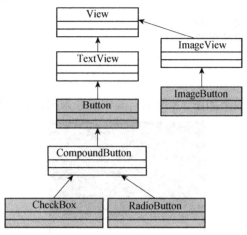

图 3.5　按钮类组件继承关系

从图 3.5 中可以看出，Button 组件继承自 TextView 组件，而 ImageButton 组件继承自 ImageView 组件，虽然这两个组件的直接父类不同，但是作用是一样的，都可以触发单击事件。而 RadioButton 组件和 CheckBox 组件都间接地继承自 Button 组件，都可以直接使用 Button 组件支持的属性和方法，所不同的是它们都比 Button 组件多了可选中的功能。下面将对这四种按钮类组件分别进行介绍。

3.1.2.1　普通按钮

在 Android 手机应用程序中，按钮的应用十分广泛。在购物商城 APP 中，添加的购物车相关按钮如图 3.6 和图 3.7 所示。

图 3.6　"加入购物车"按钮

图 3.7　"去结算"按钮

在 XML 布局文件中通过<Button>标记添加普通按钮的基本语法格式如下：

```
<Button
android:id="@+id/ID 号"
android:layout_height="wrap_content"
android:layout_width="wrap_content"
android:text="显示文本"
    >
<Button>
```

登录/注册页面的布局效果如图 3.8 所示，其中的登录按钮就用到了 Button 组件。

图 3.8　登录/注册页面布局效果

布局中"登录"按钮组件对应的代码如下：

```
<Button
android:id="@+id/mr_login"
android:layout_width="match_parent"
android:layout_height="@dimen/dp_40"
android:background="@color/red"
android:textColor="@color/white"
android:layout_marginTop="@dimen/dp_16"
android:textSize="@dimen/sp_20"
android:text="登录" />
```

在屏幕上添加按钮后，只有为按钮添加点击事件监听器，才能使按钮发挥其特有的作用。Android 系统提供了四种为按钮添加单击事件监听器的方法。

（1）使用匿名内部类进行监听，为每个控件都设置一个监听器，当控件较少时使用这种方法比较好。具体方法如下：

```
private Button button1;
button1 = (Button) findViewById(R.id.button1);
        // 方法一
        button1.setOnClickListener(new OnClickListener() {

            @Override
            public void onClick(View v) {
                Toast.makeText(MainActivity.this, "按钮 1 被点击了", Toast.LENGTH_SHORT).show();
```

            }
        });

(2) 在 MainActivity 类中实现 OnClickListener 接口，重写该接口中的 onClick 方法，多个控件对应同一个监听事件，在回调方法中使用 case 语句分别进行处理。该方法适用于按钮较多的时候。具体方法如下：

```java
public class MainActivity extends AppCompatActivity implement View.OnClickListener{
    private Button button2;
    private Button button3;
    @Override
    protected void onCreate(Bundle savedInstanceState) {
        super.onCreate(savedInstanceState);
        setContentView(R.layout.activity_main);

        button2 = (Button) findViewById(R.id.button2);
        button3 = (Button) findViewById(R.id.button3);

        // 方法二
        button2.setOnClickListener(this);
        button3.setOnClickListener(this);
    }

    @Override
    public void onClick(View v) {
        switch (v.getId()) {
            case R.id.button2:
                Toast.makeText(MainActivity.this, "按钮 2 被点击了", Toast.LENGTH_SHORT).show();
                break;
            case R.id.button3:
                Toast.makeText(MainActivity.this, "按钮 3 被点击了", Toast.LENGTH_SHORT).show();
                break;
        }
    }
}
```

(3) 在 XML 文件中为按钮添加 onClick 属性，然后在 .java 文件中编写一个以 onClick 的属性值为方法名的方法。该方法多用于做测试，建议尽量不要使用这种方法。具体方法如下：

```xml
<Button
    android:id="@+id/button4"
    android:layout_width="wrap_content"
    android:layout_height="wrap_content"
    android:onClick="onClick"
    android:text="按钮监听方法三" />
```

```java
public void onClick(View view) {
    Toast.makeText(MainActivity.this, "按钮 4 被点击了", Toast.LENGTH_SHORT).show();
}
```

(4) 用于多个按钮对应一个监听事件的情况，也可以用为多个按钮设置同一个监听器对象的方式来实现。具体方法如下：

```java
private Button button1;
    private Button button2;

    @Override
    protected void onCreate(Bundle savedInstanceState) {
```

```java
        super.onCreate(savedInstanceState);
        setContentView(R.layout.activity_main);

        button1 = (Button) findViewById(R.id.button1);
        button2 = (Button) findViewById(R.id.button2);
        button1.setOnClickListener(listener);
        button2.setOnClickListener(listener);
    }
    // 方法四
    View.OnClickListener listener = new View.OnClickListener() {

        @Override
        public void onClick(View v) {
            switch (v.getId()) {
            case R.id.button1:
                Toast.makeText(MainActivity.this, "按钮 1 被点击了", Toast.LENGTH_SHORT).show();
                break;
            case R.id.button2:
                Toast.makeText(MainActivity.this, "按钮 2 被点击了", Toast.LENGTH_SHORT).show();
                break;
            }
        }
    };
```

3.1.2.2 图片按钮

图片按钮与普通按钮的使用方法基本相同，只不过图片按钮使用标记定义，并且可以为其指定 android.src 属性，用于设置要显示的图片。在布局文件中添加图片按钮的基本语法格式如下：

```xml
<ImageButton
    android:id="@+id/ID 号"
    android:layout_height="wrap_content"
    android:layout_width="wrap_content"
    android:src="@mipmap/图片文件名"
    android:scaleType="缩放方式"
    >
</ImageButton>
```

重要属性说明如下。

（1）android:src 属性：用于指定按钮上显示的图片。

（2）android:scaleType 属性：用于指定图片的缩放方式，具体如表 3.2 所示。

表 3.2 图片缩放方式

方 式	描 述
center	在视图中心显示图片，并且不缩放图片
centerCrop	按比例缩放图片，使图片长（宽）大于等于视图的相应维度
centerInside	按比例缩放图片，使图片长（宽）小于等于视图的相应维度
fitCenter	按比例缩放图片，使图片显示在视图内部，且居中显示
fitEnd	按比例缩放图片，使图片显示在视图内部，且显示在视图的下半部分位置
fitStart	按比例缩放图片，使图片显示在视图内部，且显示在视图的上半部分位置
fitXY	把图片不按比例缩放到视图的大小，并显示
matrix	用矩阵绘制图片

3.1.2.3 单选按钮

在默认情况下，单选按钮显示为一个圆形图标，并且在该图标旁边放置一些说明性文字。在程序中，一般将多个单选按钮放置在按钮组中，使这些单选按钮表现出类似单选的功能，即当用户选中某个单选按钮后，按钮组中的其他按钮将被自动取消选中状态。

在布局文件中添加 RadioButton 组件的基本语法格式如下：

```
<RadioButton
android:text="显示文本"
android:id="@+id/ID 号"
android:checked="true|false"
android:layout_width="wrap_content"
android:layout_height="wrap_content"
>
</RadioButton>
```

RadioButton 组件的 android:checked 属性用于指定选中状态，属性值为 True 时表示选中；属性值为 False 时表示取消选中，默认为 False。

通常情况下，RadioButton 组件需要与 RadioGroup 组件一起使用，组成一个单选按钮组。在 XML 布局文件中，添加 RadioGroup 组件的基本语法格式如下：

```
<RadioGroup
    android:id="@+id/ID 号"
    android:orientation="horizontal"
    android:layout_width="warp_content"
    android:layout_height="wrap_content">
<!--添加多个 RadioButton 组件-->
</RadioGroup>
```

例如，在页面中添加一个选择性别的单选按钮组和一个"提交"按钮，添加按钮后的效果如图 3.9 所示，具体代码如下：

```
<RadioGroup android:id="@+id/radioGroup1"
android:layout_width="wrap_content"
android:layout_height="wrap_content"
android:orientation="horizontal">
<RadioButton android:id="@+id/radio0"
android:layout_width="wrap_content"
android:layout_height="wrap_content"
android:text="男"
android:checked="true"/>
<RadioButton android:id="@+id/radio1"
android:layout_width="wrap_content"
android:layout_height="wrap_content"
android:text="女"/>
</RadioGroup>
<Button android:text="提交"
android:id="@+id/button1"
android:layout_width="wrap_content"
android:layout_height="wrap_content"/>
```

在屏幕中添加单选按钮组后，还需要获取单选按钮组中选中项的值，有以下两种获取方法。

1. 在改变单选按钮组的值时获取

获取选中的单选按钮的值,首先需要获取单选按钮组,然后为其添加 OnCheckedChange 事件监听器,并在其 onCheckedChanged 方法中根据参数 checkedId 获取被选中的单选按钮,并通过其 getText 方法获取该单选按钮对应的值。例如,若要获取 ID 属性为 radioGroup1 的单选按钮组的值,则可以通过以下代码实现:

图 3.9 添加按钮后的效果

```
RadioGroup sex=(RadioGroup)findViewById(R.id.radioGroup1);
    sex.setOnCheckedChangeListener(new OnCheckedChangeListener() {

        @Override
        public void onCheckedChanged(RadioGroup group, int checkedId) {
            RadioButton rdo=(RadioButton)findViewById(checkedId);
            rdo.getText();//获取被选中的单选按钮的值
        }
    });
```

2. 在单击"提交"按钮(或其他按钮)时获取值

在该按钮的单击事件监听器的 onClick 方法中,通过 for 循环语句遍历当前单选按钮组,并根据被遍历的单选按钮的 isChecked 方法判断按钮是否被选中,当被选中时,通过单选按钮的 getText 方法获取对应的值,具体代码如下:

```
final RadioGroup sex=(RadioGroup)findViewById(R.id.radioGroup1);
    Button button=(Button)findViewById(R.id.button1);
    button.setOnClickListener(new OnClickListener() {

        @Override
        public void onClick(View arg0) {
            for (int i = 0; i < sex.getChildCount(); i++) {
                RadioButton r=(RadioButton)sex.getChildAt(i);
                if(r.isChecked()){
                    r.getText();
                    break;
                }
            }
        }
    });
```

3.1.2.4 复选框按钮

在默认情况下,复选框按钮显示为一个方块图标,并且在该图标旁边放置一些说明性文字。与单选按钮唯一不同的是,复选框按钮可以进行多选设置,每个复选框按钮都提供"选中"和"不选中"两种状态。

在 XML 布局文件中通过<CheckBox>添加复选框按钮的基本语法格式如下:

```
< CheckBox android:text="显示文本"
android:id="@+id/ID 号"
android:layout_width="wrap_content"
android:layout_height="wrap_content"
    >
</ CheckBox >
```

由于使用复选框按钮可以选中多项，所以为了确定用户是否选中了某一项，还需要为每个选项都添加事件监听器。例如，为名称为 check1 的复选框按钮添加状态改变事件监听器，可以使用下面的代码：

```
final CheckBox    check1=(CheckBox)findViewById(R.id. check1);
check1.setOnCheckedChangeListener(new CompoundButton.OnCheckedChangeListener(){
                @Override
                public void onCheckedChanged(CompoundButton buttonView,
                        boolean isChecked) {
                    if(isChecked){
check1.getText();//选中
                    }else{
//未选中
                    }
                }
            });
```

3.1.3 图像类组件

Android 系统提供了比较丰富的图像类组件。用于显示图像的组件称为图像视图，用 ImageView 表示；用于按照行、列的方式来显示多个元素（如图像、文字等）的组件称为网格视图，用 GridView 表示。它们之间的继承关系如图 3.10 所示。

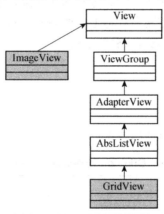

图 3.10　图像类组件继承关系

从图 3.10 中可以看出，ImageView 组件继承自 View，主要用于显示图像；GridView 组件间接地继承自 AdapterView 组件，可以包括多个列表项，并可以通过合适的方式显示。

3.1.3.1　图像视图

图像视图（ImageView）用于在屏幕中显示任何 Drawable 对象，通常用来显示图像。

在使用 ImageView 组件显示图像时，通常要将显示的图像放置在 res\drawabIe 或 res\mipmap 目录中。

在布局文件中添加图像视图，可以使用标记来实现，具体的语法格式如下：

```
<ImageView
    属性列表
    >
```

</ ImageView >

ImageView 组件支持的常用 XML 属性如表 3.3 所示。

表 3.3 ImageView 组件支持的常用 XML 属性

XML属性	描　述
android:adjustViewBounds	设置ImageView是否调整自己的边界来保持所显示图片的长宽比
android:maxHeight	设置ImageView的最大高度，需要设置android:adjustViewBounds属性值为True，否则不起作用
android:maxWidth	设置ImageView的最大宽度，需要设置android:adjustViewBounds属性值为True，否则不起作用
android:scaleType	设置图片的填充方式。 matrix：用矩阵绘图； fitXY：拉伸图片（不按比例）以填充View的宽/高； fitStart：按比例拉伸图片，拉伸后图片的高度为View的高度，且显示在View的左侧； fitCenter：按比例拉伸图片，拉伸后图片的高度为View的高度，且显示在View的中间； fitEnd：按比例拉伸图片，拉伸后图片的高度为View的高度，且显示在View的右侧； center：按原图大小显示图片，但图片宽/高大于View的宽/高时，仅截取图片中间部分显示
android:src	设置View的drawable资源

3.1.3.2 网格视图

网格视图（GridView）是指按照行、列分布的方式来显示多个组件，通常用于显示图像或图标等。在使用网格视图时，需要在屏幕上添加 GridView 组件，通常在 XML 布局文件中使用<GridView>标记实现，其基本语法格式如下：

```
< GridView
    属性列表
    >
</ GridView >
```

GridView 组件支持的 XML 属性如表 3.4 所示。

表 3.4 GridView 组件支持的 XML 属性

XML属性	描　述
android:columnWidth	设置列的宽度
android:gravity	设置对齐方式
android:horizontalSpacing	设置各元素之间的水平间距
android:numColumns	设置列数，属性值大于1
android:stretchMode	设置GridView中的条目以何种缩放方式填充剩余空间

在使用 GndView 组件时，通常使用 Adapter 类为 GridView 组件提供数据。

Adapter 类是一个接口，代表适配器对象。它是组件与数据之间的桥梁，通过它可以处理数据并将其绑定到相应的组件上。实现该接口的常用类包括以下几个。

◆ ArrayAdapter：数组适配器，通常用于将数组的多个值包装成多个列表项；只能显示一行文字。

◆ SmipleAdapter：简单适配器，通常用于将 List 集合的多个值包装成多个列表项；可以自定义各种效果，其功能强大。

- SmipleCursorAdpter：与 SmipleAdapter 类似，只不过它需要将 Cursor（数据库的游标对象）的字段与组件 ID 对应，从而将数据库的内容以列表形式展示出来。
- BaseAdapter：一个抽象类，继承它需要实现较多的方法；通常它可以对各列表项进行最大限度的定制，具有很高的灵活性。

 ## 3.2 常见对话框

3.2.1 通过 Toast 类显示消息提示框

Toast 类用于在屏幕中显示一个消息提示框，该消息提示框没有任何控制按钮，并且不会获得焦点，且经过一段时间后自动消失。Toast 类通常用于显示一些快速提示信息，应用范围非常广泛。

使用 Toast 类来显示消息提示框非常简单，只需要以下步骤。

（1）创建一个 Toast 对象。通常有以下两种方法。

一种方法是使用构造方式进行创建：

```
Toast toast=new Toast(this);
```

另一种方法是调用 Toast 类的 makeText 方法创建：

```
Toast toast=Toast.makeText(this,"要显示的内容",Toast.LENGTH_SHORT);
```

（2）调用 Toast 类提供的方法来设置该消息提示框的对齐方式、页边距、显示的内容等。常用的方法如下。

- setDuration（int duration）：设置消息提示框持续的时间，参数通常使用 Toast.LENGTH_LONG 或 Toast.LENGTH_SHORT。
- setGravity(int gravity,int xOffset,int yOffset)：设置消息提示框的位置，参数 grivaty 用于指定对齐方式；xOffset 和 yOffset 用于指定具体的偏移值。
- setMargin(float horizontalMargin,float verticalMargin)：设置消息提示的页边距。
- setText(CharSequence s)：设置要显示的文本内容。
- setView(View view)：设置要在提示框中显示的视图。

3.2.2 使用 AlertDialog 类实现对话框

AlertDialog 类的功能非常强大，它不仅可以生成带按钮的提示对话框，还可以生成带列表的列表对话框。

使用 AlertDialog 类生成的对话框通常可分为四个区域，分别是图标区、标题区、内容区和按钮区。例如，如图 3.11 所示的提示对话框可分为四个区域。

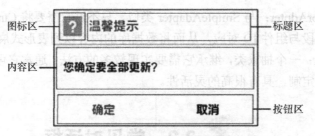

图 3.11 提示对话框的四个区域

概括起来，使用 AlertDialog 类可以生成的对话框有以下四种。
- 带确定、中立或取消等多个按钮的提示对话框，其按钮个数不是固定的，可以根据需要添加。例如，如果不需要中立按钮，那么就可以生成只带有确定和取消按钮的提示对话框，也可以是只带有一个按钮的提示对话框。
- 带列表的列表对话框。
- 带多个单选列表项和多个按钮的列表对话框。
- 带多个多选列表项和多个按钮的列表对话框。

在使用 AlertDialog 类生成对话框时的常用方法如表 3.5 所示。

表 3.5 AlertDialog 类常用方法

方法	描述
setTitle(CharSequence)	设置标题
setContentText(CharSequence)	设置内容
setIcon(Drawable icon)	设置显示的图标
setIcon(int iconId)	设置要显示图标的资源 ID
setMessage(CharSequence message)	设置显示信息的字符串

通常情况下，使用 AlertDialog 类只能生成带多个按钮的提示对话框，若要生成另外三种列表对话框，则需要使用 AlertDialog.Builder 类。AlertDialog.Builder 类常用方法如表 3.6 所示。

表 3.6 AlertDialog.Builder 类常用方法

方法	描述
setTitle(CharSequence)	设置标题
setContentText(CharSequence)	设置内容
setIcon(Drawable icon)	设置显示的图标
setIcon(int iconId)	设置要显示图标的资源 ID
setMessage(CharSequence message)	设置显示信息的字符串
setNeutralButton	设置普通按钮
setPositiveButton	为对话框添加"Yes"按钮
setNegativeButton	为对话框添加"No"按钮
create	创建对话框
show	显示对话框

3.2.3 使用 Notification 类在状态栏上显示通知

状态栏位于手机屏幕的最上方，通常用于显示手机当前的网络状态、系统时间及电池状态等信息。在使用手机时，当有未接来电或有新的短消息时，手机会给出相应的提示信息，这些提示信息通常会显示在手机屏幕的状态栏上。

Android 系统提供了用于处理这些信息的类，它们是 Notification 类和 NotificationManager 类。其中，Notificafion 类代表的是具有全局效果的通知，而 NotificationManager 类则用来发送 Notification 通知的系统服务。

使用 Notification 类和 NotificationManager 类发送和显示通知的方法比较简单，大致有以下四个步骤。

（1）调用 getSystem Service 方法用于获取系统的 NotificationManager 服务。
（2）创建一个 Notification 对象。
（3）为 Notification 对象设置各种属性，常用的方法如表 3.7 所示。
（4）创建一个 NotificationManager 对象，调用 notify 方法发送 Notificafion 通知。

表 3.7 Notification 对象中的常用方法

方　　法	描　　述
setTitle(CharSequence)	设置标题
setContentText(CharSequence)	设置内容
setSubText(CharSequence)	设置内容下面的一小行文字
setTicker(CharSequence)	设置收到通知时在顶部显示的文字信息
setWhen(long)	设置通知时间
setSmallIcon(int)	设置右下角的小图标，在收到通知时，顶部也会显示这个小图标
setLargeIcon(Bitmap)	设置左侧的大图标
setAutoCancel(Boolean)	确定用户点击 Notification 面板后是否使通知取消（默认不取消）
setDefault(int)	为通知添加声音、闪灯或震动效果

 ## 3.3 购物商城 APP 的 UI 交互

3.3.1 商城首页底部的页面选择

在购物商城 APP 商城首页的底部，需要设置四个单选按钮，用于在"首页""分类""购物车""我的"界面间进行选择。具体的实现步骤如下。

（1）将 MainActivity 继承自带有标题栏的 APPCompatActivity 类，实现 RadioGroup.On-

CheckedChangeListener 监听接口，用于实现单选按钮的选中事件监听：

```
public class MainActivity extends APPCompatActivity implements RadioGroup.OnCheckedChangeListener
```

（2）声明 radioGroup 按钮组和 fragments 碎片集合，并在 onCreate 方法中进行初始化：

```
public RadioGroup radioGroup;//底部切换按钮
private List<Fragment> fragments;//Fragment 集合
@Override
protected void onCreate(Bundle savedInstanceState) {
    super.onCreate(savedInstanceState);
    setContentView(R.layout.activity_main);

    radioGroup = (RadioGroup) findViewById(R.id.bottombar);//初始化控件
    radioGroup.setOnCheckedChangeListener(this);//设置单选按钮的选中事件监听

    fragments = new ArrayList<>();//初始化集合
}
```

（3）重写 onCheckedChanged 方法，实现单选按钮的选中事件监听回调，用于实现选择界面的功能。在该方法中，通过 switch 语句，根据单选组件的 ID 属性进行切换处理。由于界面切换主要通过 Fragment 实现，这部分内容在后面的章节将会详细介绍，因此，此处利用 Toast 弹出悬浮窗代替界面的切换。具体代码如下：

```
* 选择切换界面
*
* @param group
* @param checkedId
*/
@Override
public void onCheckedChanged(RadioGroup group, int checkedId) {

    switch (checkedId) {
        case R.id.mr_shoppingmall://商城
            Toast.makeText(MainActivity.this.getApplicationContext(),"切换至商城首页",Toast.LENGTH_SHORT).show();
            break;
        case R.id.mr_category://分类
            Toast.makeText(MainActivity.this.getApplicationContext(), "切换至商城分类列表", Toast.LENGTH_SHORT).show();
            break;
        case R.id.mr_shoppingcart://购物车
            Toast.makeText(MainActivity.this.getApplicationContext(), "切换至商城购物车", Toast.LENGTH_SHORT).show();
            break;
        case R.id.mr_mine://我的
            Toast.makeText(MainActivity.this.getApplicationContext(), "切换至商城个人中心", Toast.LENGTH_SHORT).show();
            break;
```

点击 AndroidStudio 右上方的 Run 按钮，选择模拟器运行项目，部分运行效果如图 3.12 所示。

图 3.12 商城底部切换部分运行效果

3.3.2 用户登录

（1）新建一个 empty activity，如图 3.13 所示。

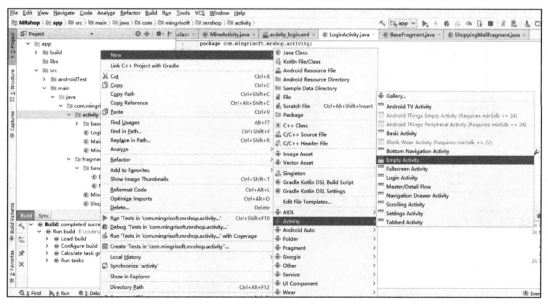

图 3.13 新建 empty activity

（2）命名为 loginActivity，参数设置如图 3.14 所示。单击"Finish"按钮，创建完成。

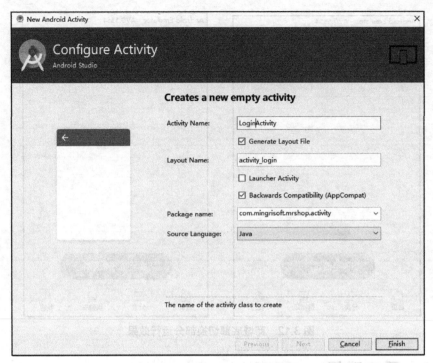

图 3.14 创建 activity 参数设置

（3）在 AndroidManifest.xml 配置文件中设置 LoginActivity 为默认启动，如图 3.15 所示。

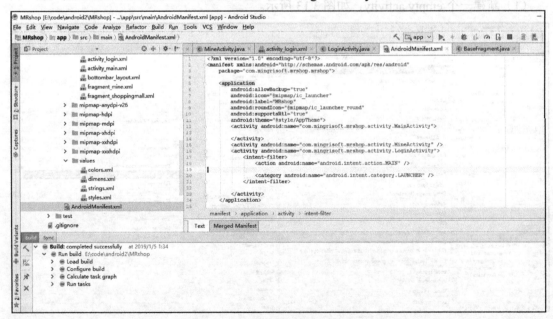

图 3.15 AndroidManifest.xml 配置文件

采用线性布局管理器实现登录/注册页面。登录/注册页面布局主要包含两个 TextView 控件、一个 EditText 和一个 Button 按钮，如图 3.16 所示。

图 3.16　登录/注册页面布局

具体代码如下：

```xml
<?xml version="1.0" encoding="utf-8"?>
<LinearLayout xmlns:android="http://schemas.android.com/apk/res/android"
    android:orientation="vertical" android:layout_width="match_parent"
    android:background="@color/white"
    android:padding="@dimen/dp_16"
    android:layout_height="match_parent">

    <TextView
        android:id="@+id/mr_login_title"
        android:layout_width="match_parent"
        android:layout_height="wrap_content"
        android:gravity="center"
        android:layout_marginBottom="@dimen/dp_16"
        android:textSize="@dimen/sp_20"
        android:text="登录" />

    <EditText
        android:id="@+id/mr_user"
        android:layout_width="match_parent"
        android:layout_height="wrap_content"
        style="@style/Mine_Text_Style"
        android:layout_marginTop="@dimen/dp_16"
        android:hint="请输入账号"
        android:drawableLeft="@drawable/user"
        android:drawablePadding="@dimen/dp_10"
        android:textSize="@dimen/sp_20"
        android:background="@drawable/line_edittext"
        android:lines="1"
        />

    <EditText
        android:id="@+id/mr_password"
        android:layout_width="match_parent"
        android:layout_height="wrap_content"
        style="@style/Mine_Text_Style"
```

```xml
        android:layout_marginTop="@dimen/dp_16"
        android:drawableLeft="@drawable/pwd"
        android:drawablePadding ="@dimen/dp_10"
        android:background="@drawable/line_edittext"
        android:inputType="textPassword"
        android:hint="请输入密码"
        android:lines="1"
        android:textSize="@dimen/sp_20"
        />

    <Button
        android:id="@+id/mr_login"
        android:layout_width="match_parent"
        android:layout_height="@dimen/dp_40"
        android:background="@drawable/add_thing_btn_selector"
        android:textColor="@color/white"
        android:layout_marginTop="@dimen/dp_16"
        android:textSize="@dimen/sp_20"
        android:text="登录" />

    <TextView
        android:id="@+id/mr_reg"
        android:layout_width="match_parent"
        android:layout_height="wrap_content"
        android:layout_marginTop="@dimen/dp_5"
        android:text="马上注册" />
</LinearLayout>
```

后台代码实现步骤如下。

(1) 后台的 LogActivity 继承自 APPCompatActivity，实现 View.OnClickListener 监听。首先声明所需组件，然后初始化各组件，并赋予登录按钮和注册文本点击事件：

```java
public class LoginActivity extends APPCompatActivity implements View.OnClickListener {
    private TextView et_custom_title;   //我的页面标题
    private TextView et_title;          //标题
    private TextView et_reg;            //注册
    private EditText et_user;           //账号
    private EditText et_pwd;            //密码
    private Button btn_login;           //登录按钮
    private String user;                //输入的账号
    private String pwd;                 //输入的密码

    @Override
    protected void onCreate(Bundle savedInstanceState) {
        super.onCreate(savedInstanceState);
        setContentView(R.layout.activity_login);

        //初始化按钮
        et_title = (TextView) findViewById(R.id.mr_login_title);
        et_custom_title= (TextView) findViewById(R.id.custom_login);
        et_reg = (TextView) findViewById(R.id.mr_reg);
        et_reg.setClickable(true);
        et_reg.setOnClickListener((View.OnClickListener) this);
        et_user = (EditText) findViewById(R.id.mr_user);
        et_pwd = (EditText) findViewById(R.id.mr_password);
        btn_login = (Button) findViewById(R.id.mr_login);
```

```
        btn_login.setOnClickListener(this);
    }
```

（2）重写点击事件的回调方法 onClick，以实现：点击登录按钮，模拟登录成功；点击注册文本，模拟跳转注册页面。

```
@Override
public void onClick(View view) {
    user=et_user.getText().toString();
    pwd=et_pwd.getText().toString();
    switch (view.getId()) {
        case R.id.mr_login://登录注册文本框
            if("".equals(user)||user==null||"".equals(pwd)||pwd==null) {
                Toast.makeText(LoginActivity.this.getApplicationContext(), "登录失败，账号或密码不能为空", Toast.LENGTH_LONG).show();
            }else{
                Toast.makeText(LoginActivity.this.getApplicationContext(), "登录成功！账号为"+user+",密码为"+pwd, Toast.LENGTH_LONG).show();
            }
            break;
        case R.id.mr_reg:
            Toast.makeText(LoginActivity.this.getApplicationContext(), "切换至注册页面", Toast.LENGTH_SHORT).show();
            Intent intent=new Intent(LoginActivity.this, RegisterActivity.class);
            startActivity(intent);
            break;
    }
}
```

⚠ 注：在下一章会详细介绍注册成功和注册所涉及的界面跳转。此处以弹出悬浮窗作为模拟。

3.3.3 用户注册

（1）新建 RegisterActivity，并在 AndroidManifest.xml 配置文件中注册该 Activity：

```
<activity android:name="com.mingrisoft.mrshop.activity.RegisterActivity">
```

（2）在 activity_register.xml 布局文件中修改布局，分别添加五个 EditText 和一个 TextView 等文本类组件，以及一个 RadioGroup 组件、一个 Button 和一组 RadioGroup 等按钮类组件，具体代码如下：

```
<?xml version="1.0" encoding="utf-8"?>
<LinearLayout xmlns:android="http://schemas.android.com/apk/res/android"
    android:orientation="vertical" android:layout_width="match_parent"
    android:background="@color/white"
    android:padding="@dimen/dp_16"
    android:layout_height="match_parent">

    <EditText
        android:id="@+id/editText_zcyh"
        style="@style/Mine_Text_Style"
        android:layout_width="match_parent"
        android:layout_height="wrap_content"

        android:background="@drawable/line_edittext"
```

```xml
        android:drawableLeft="@drawable/pwd"
        android:drawablePadding="@dimen/dp_10"
        android:hint="请输入用户名"
        android:lines="1"
        android:textSize="@dimen/sp_20" />

    <EditText
        android:id="@+id/editText_mia"
        android:layout_width="match_parent"
        android:layout_height="wrap_content"
        style="@style/Mine_Text_Style"

        android:drawableLeft="@drawable/pwd"
        android:drawablePadding ="@dimen/dp_10"
        android:background="@drawable/line_edittext"
        android:inputType="textPassword"
        android:hint="请输入密码"
        android:lines="1"
        android:textSize="@dimen/sp_20"
        />
    <EditText
        android:id="@+id/editText_qrmia"
        android:layout_width="match_parent"
        android:layout_height="wrap_content"
        style="@style/Mine_Text_Style"

        android:drawableLeft="@drawable/pwd"
        android:drawablePadding ="@dimen/dp_10"
        android:background="@drawable/line_edittext"
        android:inputType="textPassword"
        android:hint="请确认密码"
        android:lines="1"
        android:textSize="@dimen/sp_20"
        />
    <EditText
        android:id="@+id/editText_sjh"
        android:layout_width="match_parent"
        android:layout_height="wrap_content"
        style="@style/Mine_Text_Style"

        android:drawableLeft="@drawable/pwd"
        android:drawablePadding ="@dimen/dp_10"
        android:background="@drawable/line_edittext"
        android:hint="请输入手机号码"
        android:lines="1"
        android:textSize="@dimen/sp_20"
        />
    <EditText
        android:id="@+id/editText_dzyx"
        android:layout_width="match_parent"
        android:layout_height="wrap_content"
        style="@style/Mine_Text_Style"

        android:drawableLeft="@drawable/pwd"
        android:drawablePadding ="@dimen/dp_10"
        android:background="@drawable/line_edittext"
        android:hint="请输入电子邮箱"
        android:lines="1"
        android:textSize="@dimen/sp_20"
        />
```

```xml
<RadioGroup
    android:id="@+id/group_sex"
    style="@style/MineBottomBarStyle"
    android:layout_height="wrap_content">

    <RadioButton
        android:id="@+id/rd_man"
        style=""
        android:checked="true"
        android:text="男" />

    <RadioButton
        android:id="@+id/rd_woman"
        style=""
        android:text="女" />
</RadioGroup>
<Button
    android:id="@+id/button_qd"
    android:layout_width="fill_parent"
    android:layout_height="26dp"
    android:layout_weight="1"
    android:background="@drawable/add_thing_btn_selector"
    android:text="确定"
    android:textColor="@color/white"
    android:textSize="15dp" />

<TextView
    android:id="@+id/mr_reg"
    android:layout_width="match_parent"
    android:layout_height="wrap_content"
    android:layout_marginTop="@dimen/dp_5"
    android:text="返回登录" />
</LinearLayout>
```

运行项目，效果如图3.17所示。

图3.17 注册页面运行效果

（3）在 RegisterActivity 中重写 onCreate 方法，初始化相应控件：

```
protected void onCreate(Bundle savedInstanceState) {
    super.onCreate(savedInstanceState);
    setContentView(R.layout.activity_register);
    //初始化控件
    edt_email=(EditText)findViewById(R.id.editText_dzyx);
    edt_phone=(EditText)findViewById(R.id.editText_sjh);
    edt_usr=(EditText)findViewById(R.id.editText_zcyh);
    edt_pwd=(EditText)findViewById(R.id.editText_mia);
    edt_pwd1=(EditText)findViewById(R.id.editText_qrmia);
    radioGroup = (RadioGroup) findViewById(R.id.group_sex);

    btn_qr=(Button)findViewById(R.id.button_qd);
```

（4）为确认按钮 bt_qr 设置监听器对象，并编写回调方法，用于实现点击"确定"按钮后判断密码与确认密码是否一致，以及前几个输入项（用户名、密码、确认密码）是否不为空，若判断为真，则弹出悬浮窗，并显示提示信息"恭喜！注册成功！"，然后返回用户登录界面；否则所有输入框置空，弹出悬浮窗，并显示提示信息"输入有误，请重新输入！"。具体代码如下：

```
btn_qr.setOnClickListener(new View.OnClickListener() {
    @Override
    public void onClick(View view) {
        //点击"确定"按钮后，判断密码与确认密码是否一致，以及前几个输入项（用户名、密码、
确认密码）是否不为空，若判断为真，则弹出悬浮窗，并显示提示信息"恭喜！注册成功！"并结束当前界
面（返回用户登录界面）；否则所有输入框置空，弹出悬浮窗，并显示提示信息"输入有误，请重新输入！"
        if(edt_pwd.getText().toString().trim().equals(edt_pwd1.getText().toString().trim())
                && !edt_usr.getText().toString().trim().equals("") ){
            radio_sex = (RadioButton)findViewById(radioGroup.getCheckedRadioButtonId());
            //输入的信息有效，注册成功
            Toast.makeText(RegisterActivity.this,"恭喜！注册成功！",Toast.LENGTH_SHORT).show();
            Intent intent=new Intent(RegisterActivity.this, LoginActivity.class);
startActivity(intent);
        }else {
            Toast.makeText(RegisterActivity.this,"输入有误，请重新输入！",Toast.LENGTH_SHORT).show();
        }
    }
});
```

⚠ 注：在下一章节会详细介绍注册成功和注册所涉及的界面跳转。此处以弹出悬浮窗作为模拟。

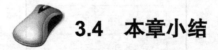 **3.4 本章小结**

本章主要介绍 Android 应用开发中常用的文本类组件、按钮类组件及图像类组件。这些组件都属于 UI 组件，可以在界面中看到效果。其中，文本类组件用于输入和输出文字，按钮类组件用于触发一定的事件，图像类组件用于显示图像。本章最后还介绍了用户登录注册页

面的 UI 交互设计。在实现 UI 界面交互时，经常会应用到这些组件，需要读者认真学习，灵活运用。

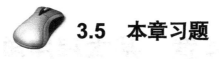

3.5 本章习题

普通按钮与图片按钮的区别是什么？

第 4 章　基本程序单元 Activity

在前面介绍的实例中已经应用过 Activity，不过那些实例中的所有操作都是在一个 Activity 中进行的。在实际的应用开发中，经常需要包含多个 Activity，而且这些 Activity 之间可以相互跳转或传递数据。本章对 Activity 进行详细介绍。

4.1　Activity 概述

Android 应用程序的开发过程中经常用到四个基本组件，分别是活动（Activity）、服务（Service）、广播接收者（BroadcastReceiver）和内容提供者（ContentProvier）。其中，Activity 是 Android 应用程序中最常见的组件之一。在 Android 应用程序中，布局文件代表了用户看到的前端可视化界面，Activity 则代表了可视化界面后台实现的功能。换句话说，布局和 Activity 共同实现了用户在 APP 上进行操作的人机交互功能。

在一个 Android 应用程序中可以有多个 Activity，这些 Activity 组成了 Activity 栈（Stack），当前活动的 Activity 位于栈顶，之前活动的 Activity 被压至下面，成为非活动 Activity，等待被恢复为活动状态。在 Activity 的生命周期中有四个重要状态，如表 4.1 所示。

表 4.1　Activity 的四个重要状态

状　　态	描　　述
运行状态	当 Activity 位于栈顶时，活动就处于运行状态，且用户可见。被系统回收可能性最小的就是运行状态的活动
暂停状态	失去焦点的 Activity，仍然用户可见，但是在内存小的情况下，不能被系统回收
停止状态	当 Activity 不处于栈顶位置，且用户安全不可见时，就进入停止状态，当内存较小时系统会回收这样的活动
销毁状态	该 Activity 结束，或者 Activity 所在的虚拟器进程结束

在了解了 Activity 的四个重要状态后，再来看如图 4.1 所示的 Activity 生命周期。该图显示了 Activity 生命周期中的各种重要状态，以及相关的回调方法（图中的"活动"指 Activity）。

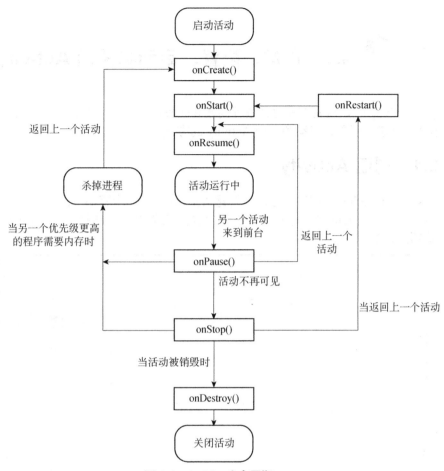

图 4.1 Activity 生命周期

在图 4.1 中，矩形框中的内容表示的内容为可以被回调的方法，而椭圆形则表示 Activity 的重要状态。从图 4.1 中可以看出，在一个 Activity 的生命周期中有一些方法会被系统回调，这些方法及其描述如表 4.2 所示。

表 4.2 Activity 生命周期中被系统回调的方法

方法	描述
onCreate()	该方法在 Activity 第一次启动时被调用，在该方法中初始化 Activity 所能使用的全局资源和状态，如绑定事件、创建线程等
onStart()	该方法当 Activity 对用户可见时被调用，即 Activity 展现在前端。该方法一般用来初始化或启动与更新界面相关的资源
onResume()	该方法当用户与 Activity 进行交互时被调用，此时 Activity 位于返回栈的栈顶，并处于运行状态。该方法完成一些轻量级的工作，以避免用户等待
onPause()	该方法在启动或恢复另一个 Activity 时被调用。该方法一般用来保存界面的持久信息，提交未保存的数据，并释放消耗 CPU 的资源
onStop()	该方法在 Activity 为不可见状态时被调用，如其他 Activity 启动或恢复并将其覆盖时
onDestroy()	该方法在 Activity 销毁之前被调用
onRestart()	该方法当 Activity 重新启动时被调用

4.2 创建、配置、启动和关闭 Activity

在使用 Activity 时，需要首先对其进行创建和配置，然后才可以启动或关闭 Activity。下面将详细介绍创建、配置、启动和关闭 Activity 的方法。

4.2.1 创建 Activity

使用 Android Studio 也可以很方便地创建 Activity，具体步骤如下。

（1）在 APP 节点上单击鼠标右键，在弹出快捷菜单中依次选择 New→Activity→Empty Activity 选项，如图 4.2 所示。

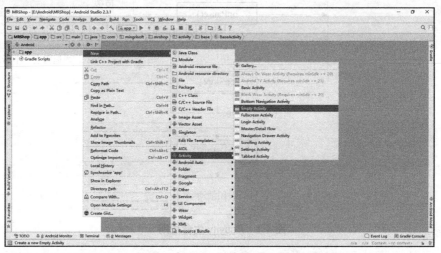

图 4.2 创建 Empty Activity

（2）在弹出的对话框中修改 Activity 的名称，如图 4.3 所示。

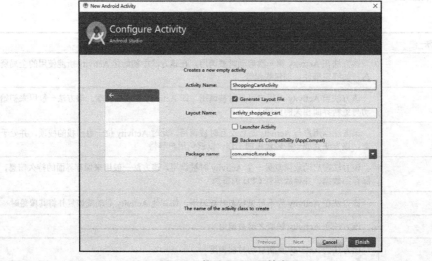

图 4.3 修改 Activity 的名称

（3）单击"Finish"按钮，创建一个空的 Activity，以及与之对应的布局文件。

4.2.2 配置 Activity

使用 Android Studio 向导创建 Activity 后，会自动在 AndroidManifest.xml 文件中配置与注册该 Activity：

```
<activity android:name=".MainActivity">
    <intent-filter>
        <action android:name="android.intent.action.MAIN" />

        <category android:name="android.intent.category.LAUNCHER" />
    </intent-filter>
</activity>
```

配置 Activity 的基本语法格式如下：

```
<activity android:name="实现类"： 指定对应的 Activity 实现类
android:lable="说明性文字"  ： 指定标签
android:theme="主题"       ： 设置要应用的主题
…
>
<!—如果有多个Activity，则设置默认Activity 的方式是在AndroidManifest.xml 中为Activity 添加intent-filter，并将action 设置为android.intent.action.MAIN，将category 设置为android.intent.category.LAUNCHE-->
<intent-filter>
        <action android:name="android.intent.action.MAIN" />
        <category android:name="android.intent.category.LAUNCHER" />
    </intent-filter>
</activity>
```

4.2.3 启动和关闭 Activity

1. 启动 Activity

启动 Activity 有以下两种方式。
- 当一个 Android 应用程序中只存在一个 Activity 时，只需要在 AndroidManifest.xml 文件中对其进行配置，并且将其设置为程序的入口，这样，当运行该项目时，系统将自动启动该 Activity。
- 当一个 Android 应用程序中存在多个 Activity 时，只需要应用 startActivity 方法来启动需要的 Activity。startActivity 方法的语法格式如下：

```
public void startActivity （ Intent  intent ）
```

该方法没有返回值，只有一个 Intent 类型的入口参数。Intent 是 Android 应用程序里各组件之间的通信方式，一个 Activity 通过 Intent 表达自己的"意图"。在创建 Intent 对象时，需要指定想要被启动的 Activity。

 说明：关于 Intent 的详细介绍请参见本书 4.2.4 节。

例如，启动一个名称为"ShoppingCartActivity"的 Activity，可以使用以下代码：

```
Intent intent=new Intent(MainActivity.this, ShoppingCartActivity.class);
startActivity(intent);
```

2. 关闭 Activity

在 Android 系统中，如果想要关闭当前的 Activity，则可以使用 Activity 类提供的 finish 方法。finish 方法的语法格式如下：

```
Public void finish()
```

该方法的使用比较简单，既没有入口参数，也没有返回值，只需要在 Activity 相应的事件中调用该方法即可。

4.2.4 Intent 介绍

Intent 中文意思为"意图"。它是 Android 应用程序中传输数据的核心对象，在 Android 系统官方文档中，Intent 的定义是执行某项操作的一个抽象描述。它可以开启新的 Activity，也可以发送广播消息，或者开启 Service 服务。下面将对 Intent 及其基本应用分别进行介绍。

一个 Android 应用程序主要由 Activity、Service 和 BroadcastReceiver 三种组件组成，这三种组件是相互独立的，它们之间可以互相调用、协调工作，最终组成一个真正的 Android 应用程序。这些组件之间的通信主要由 Intent 协助完成。Intent 负责对应用中一次操作的 Action（动作）涉及的 Data（数据）、Extras（附加数据）进行描述，Android 应用程序则根据 Intent 的描述找到相应的组件，将 Intent 传递给调用的组件并完成组件的调用。因此，Intent 起着一个媒体中介的作用，专门提供组件之间互相调用的相关信息，实现调用者与被调用者之间的解耦。

4.2.5 显式 Intent 和隐式 Intent

4.2.5.1 显式 Intent

显式 Intent 是指在创建 Intent 对象时就指定接收者（如 Activity、Service 或 BroadcastReceiver），这是因为已经知道要启动的 Activity 或 Service 类的名称。这里以 Activity 为例介绍如何使用显式 Intent。

在启动 Activity 时必须在 Intent 中指明要启动的 Activity 所在的类。通常情况下，在一个 Android 项目中，如果只有一个 Activity，那么只需要在 AndroidManifest.xml 文件中进行配置，并且将其设置为程序的入口。这样，当运行该项目时系统将自动启动该 Activity。否则，就需要应用 Intent 和 startActivity 方法来启动需要的 Activity，即通过显式 Intent 来启动，具体步骤如下。

（1）创建 Intent 对象，可以使用下面的语法格式：

```
Intent intent=new Intent(Context packageContext, Class<?> cls)
```

- ◆ intent：用于指定对象名称。
- ◆ packageContext：用于指定一个启动 Activity 的上下文对象，如 MainActivity.this。
- ◆ cls：用于指定要启动的 Activity 所在的类，如 ShoppingCartActivity.class。

例如，创建一个启动 ShoppingCartActivity 的 Intent 对象，可使用下面的代码：

```
Intent intent=new Intent(MainActivity.this, ShoppingCartActivity.class);
```

（2）创建 Intent 对象后用 startActivity 方法来启动，具体代码如下：

```
startActivity(intent);
```

4.2.5.2 隐式 Intent

隐式 Intent 是指在创建 Intent 对象时不指定具体的接收者，而是定义要执行的 Action、Category 和 Data 属性，然后使 Android 系统根据相应的匹配机制找到要启动的 Activity。

使用隐式 Intent 启动 Activity 时，需要为 Intent 对象定义 Action、Category 和 Data 属性，然后再调用 startActivity 方法来启动匹配的 Activity。

例如，要在自己的应用程序中展示一个网页，就可以直接调用系统中的浏览器来打开这个网页，而不必自己再编写一个浏览器。可以使用下面的语句实现：

```
Intent intent=new Intent();        //创建 Intent 对象
intent.setAction(Intent.ACTION_VIEW);   //为 Intent 设置动作
intent,setData=(Uri.parse("http://www.baidu.com"));  //为 Intent 设置数据
startActivity(intent);   //将 Intent 传递给 Activity
```

也可以使用下面的语句实现。

```
//创建 Intent 对象
Intent intent=new Intent(Intent.ACTION_VIEW, Uri.parse("http://www.baidu.com"));
//将 Intent 传递给 Activity
startActivity(intent);
```

Uri.parse()用于把字符串解释为 URL 对象，表示需要传递的数据。

Intent.ACTION_VIEW 是 Intent 的 Action，表示需要执行的动作。

Android 系统支持的标准 Action 字符串常量如表 4.3 所示。

表 4.3 标准 Action 字符串常量

Action 常量	对应字符串	简 单 说 明
ACTION_MAIN	android.intent.action.MAIN	应用程序入口
ACTION_VIEW	android.intent.action.VIEW	显示指定数据
ACTION_ATTACH_DATA	android.intent.action.ATTACH_DATA	指定某块数据将被附加到其他地方
ACTION_EDIT	android.intent.action.EDIT	编辑指定数据
ACTION_PICK	android.intent.action.PICK	从列表中选择某项数据并返回所选的数据
ACTION_CHOOSER	android.intent.action.CHOOSER	显示一个 Activity 选择器
ACTION_GET_CONTENT	android.intent.action.GET_CONTENT	使用户选择数据，并返回所选数据
ACTION_DIAL	android.intent.action.DIAL	显示拨号面板
ACTION_CALL	android.intent.action.CALL	直接向指定用户打电话
ACTION_SEND	android.intent.action.SEND	向其他人发送数据
ACTION_SENDTO	android.intent.action.SENDTO	向其他人发送消息
ACTION_ANSWER	android.intent.action.ANSWER	应答电话
ACTION_INSERT	android.intent.action.INSERT	插入数据
ACTION_DELETE	android.intent.action.DELETE	删除数据
ACTION_RUN	android.intent.action.RUN	运行维护
ACTION_SYNC	android.intent.action.SYNC	执行数据同步

Action 常量	对应字符串	简 单 说 明
ACTION_PICK_ACTIVITY	android.intent.action.PICK_ACTIVITY	用于选择 Activity
ACTION_SEARCH	android.intent.action.SEARCH	执行搜索
ACTION_WEB_SEARCH	android.intent.action.WEB_SEARCH	执行 Web 搜索
ACTION_FACTORY_TEST	android.intent.action.FACTORY_TEST	工厂测试的入口点

4.3 多个 Activity 的使用

在 Android 应用程序中经常会使用多个 Activity，且这些 Activity 之间经常需要交换数据。下面就来介绍如何使用 Bundle 在 Activity 之间交换数据，以及如何调用另一个 Activity 并返回结果。

4.3.1 使用 Bundle 在 Activity 之间交换数据

当在一个 Activity 中启动另一个 Activity 时，经常需要传递一些数据，这时就可以通过 Intent 来实现。因为 Intent 通常称为两个 Activity 之间的信使，所以将要传递的数据保存在 Intent 中，就可以将其传递到另一个 Activity 中。可以首先将要保存的数据存放在一个 Bundle 对象中，然后通过 Intent 提供的 putExtras 方法将要携带的 Bundle 对象保存到 Intent 中，最后通过执行 Intent 将数据传递出去。通过 Intent 传递数据的示意图如图 4.4 所示。

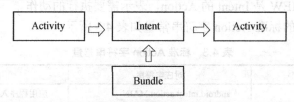

图 4.4 通过 Intent 传递数据

Bundle 是一个键值对，用于保存要传递的数据包，数据类型可以是 Boolean、Byte、Int、Long、Double 和 String 等基本类型或对应的数组，也可以是对象或对象数组。

4.3.2 调用另一个 Activity 并返回结果

在开发 Android 应用程序时，有时需要在一个 Activity 中调用另一个 Activity，当用户在第二个 Activity 结束操作后，程序将自动返回到第一个 Activity 中，第一个 Activity 能够获取并显示用户在第二个 Activity 中操作的结果。例如，程序中经常出现的"返回上一步"功能就可以通过 Intent 和 Bundle 来实现，与在两个 Activity 之间交换数据不同的是，此处需要使用 startActivityForResult 方法来启动另外一个 Activity。调用 startActivityForResult 方法启动新的 Activity 后，一旦关闭新启动的 Activity，就可以将选择的结果返回到原 Activity 中。startActivityForResult 方法的语法格式如下：

Public void startActivityForResult(Intent intent, int requestCode)

该方法将以指定的请求码启动 Activity，并且 Android 应用程序将会获取新启动的 Activity 返回的结果（通过重写 onActivityResult 方法来获取）。requestCode 参数代表启动 Activity 的请求码，该请求码的值由开发者根据需要自行设置，以用于标志请求来源。

4.4 使用 Fragment

Fragment 是 Android 3.0 新增加的概念，其中文意思是"碎片"，它与 Activity 十分相似，用来在一个 Activity 中描述一些行为或一部分用户界面。使用多个 Fragment 可以在一个单独的 Activity 中建立多个 UI 面板，也可以在多个 Activity 中重用 Fragment。例如，购物商城 APP 的商城首页就相当于一个 Activity，在这个 Activity 中包含多个 Fragment，底部四个功能界面相当于四个 Fragment，可以随意切换。商场首页底部布局如图 4.5 所示。

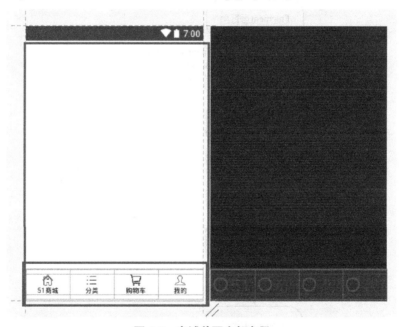

图 4.5 商城首页底部布局

4.4.1 Fragment 的生命周期

Fragment 和 Activity 一样有自己的生命周期。一个 Fragment 必须被嵌入一个 Activity，它的生命周期直接受其所属的宿主 Activity 的生命周期影响。例如，当 Activity 被暂停时，其中的所有 Fragment 也被暂停；当 Activity 被销毁时，所有隶属于它的 Fragment 也将被销毁。然而，当一个 Activity 正在运行时（处于 resumed 状态），则可以单独地对每一个 Fragment 进行操作，如添加或删除等。Fragment 的完整生命周期如图 4.6 所示。

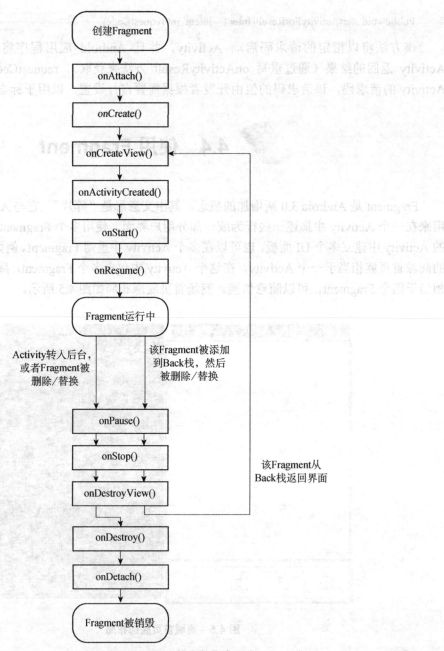

图 4.6 Fragment 的完整生命周期

4.4.2 创建 Fragment

创建一个 Fragment 前必须首先创建一个 Fragment 的子类，或者继承自另一个已经存在的 Fragment 的子类。例如，要创建一个名称为 MineFragment 的 Fragment，并重写 onCreateView 方法，可以使用以下代码：

```
public class MineFragment extends Fragment {
    @Nullable
    @Override
```

```
public View onCreateView(LayoutInflater inflater, @Nullable ViewGroup container, @Nullable Bundle savedInstanceState) {
    View view   = inflater.inflate(setLayoutID(),container,false);
    initView(view);//初始化控件
        return view;
    }
}
```

4.4.3 在 Activity 中添加 Fragment

在 Activity 中添加 Fragment 有两种方法：一种是直接在布局文件中添加，并将 Fragment 作为整个布局的一部分；另一种是在 Activity 运行时，将 Fragment 放入 Activity 对应的布局中。

1. 直接在布局文件中添加 Fragment

直接在布局文件中添加 Fragment 可以使用< fragment ></ fragment >标记实现，具体代码如下：

```
<LinearLayout xmlns:android="http://schemas.android.com/apk/res/android"
    xmlns:tools="http://schemas.android.com/tools"
    android:id="@+id/activity_main"
    android:layout_width="match_parent"
    android:layout_height="match_parent"
    android:orientation="horizontal"
    tools:context="com.example.administrator.fragmenttest.MainActivity">
    <fragment
        android:layout_width="0dp"
        android:layout_height="match_parent"
        android:layout_weight="1"
        android:name="com.example.administrator.fragmenttest.LeftFragment"
        tools:layout="@layout/left_layout" />
    <fragment
        android:layout_width="0dp"
        android:layout_height="match_parent"
        android:layout_weight="3"
        android:name="com.example.administrator.fragmenttest.RightFragment"
        tools:layout="@layout/right_layout" />
</LinearLayout>
```

2. 在 Activity 运行时添加 Fragment

在 Activity 运行时，也可以将 Fragment 添加到 Activity 布局中。实现方法是，首先获取一个 FragmentTransaction 的实例；然后使用 add 方法添加一个 Fragment，其中 add 方法的第一个参数是 Fragment 要放入的 ViewGroup（由 Resource ID 指定），第二个参数是需要添加的 Fragment；最后为了使改变生效，还必须调用 commit 方法提交事务。具体代码如下：

```
/*在 Activity 对应 Java 类中通过 getFragmentManager()
    *获得 FragmentManager，用于管理 ViewGroup 中的 Fragment
    **/
FragmentManager=getFragmentManager();
/*FragmentManager 要管理 Fragment（添加、替换及其他的执行动作）的
    *一系列的事务变化，需要通过 FragmentTransaction 来操作执行
    */
```

```
FragmentTransaction transaction = getSupportFragmentManager().beginTransaction();
    //实例化要管理的Fragment
    MineFragment mineFragment= new MineFragment();
    //通过添加（事务处理的方式）将Fragment加到对应的布局中
    transaction.add(R.id.right, mineFragment);
    //事务处理完需要提交
    transaction.commit();
```

4.5 购物商城 APP 页面的跳转和数据传递

4.5.1 商城底部的页面切换

前面介绍了商城底部页面切换的原理，本节结合本章所讲的 Fragment 真正实现商城底部的页面切换。

（1）在 MainActivity 类里新建 selectFragmentShow 方法和 addFragment 方法。

selectFragmentShow 方法用于选择要显示的 Fragment。在该方法中，首先判断 Fragment 集合中是否为空，如果为空则调用 addFragment 方法添加；然后通过 for 语句循环判断 Fragment 集合中是否有选中的元素；如果有则使其显示，并将其他的 Fragment 隐藏，否则添加并显示。

addFragment 方法用于向 Fragment 集合添加元素，在该方法中根据 FragmentTransaction 管理器和 fragment 对象，添加元素并返回 fragment 集合。具体代码如下：

```
/**
 * 选择要显示的Fragment
 *
 * @param transaction
 * @param <T>
 * @return
 */
private    <T extends Fragment> T selectFragmentShow(FragmentTransaction transaction, Class<T> clazz) {
    T fragment = null;
    boolean isHave = false;
    if (fragments.size() == 0) {//Fragment集合为空则添加
        fragment = addFragment(transaction, clazz);
        return fragment;
    }
    //判断集合中是否有选中的元素
    for (int i = 0, length = fragments.size(); i < length; i++) {
        fragment = (T) fragments.get(i);
        if (fragment.getClass().equals(clazz)) {
            transaction.show(fragment);
            isHave = true;
            continue;
        }
        transaction.hide(fragments.get(i));
    }
    if (isHave == false) {
        fragment = addFragment(transaction, clazz);
    }
```

```
        return fragment;
    }
    /**
     * 添加元素，根据FragmentTransaction 管理器和Fragment 对象，添加元素并返回Fragment 集合
     *
     * @param transaction
     * @param clazz
     * @param <T>
     * @return
     */
    public <T extends Fragment> T addFragment(FragmentTransaction transaction, Class<T> clazz) {
        T fragment = null;
        try {
            fragment = clazz.newInstance();
            transaction.add(R.id.frag_home, fragment).show(fragment);
            fragments.add(fragment);
        } catch (InstantiationException e) {
            e.printStackTrace();
        } catch (IllegalAccessException e) {
            e.printStackTrace();
        }
        return fragment;
    }
}
```

（2）在包节点上单击鼠标右键，在弹出的快捷菜单中依次选择 New→Package，新建 Fragment 包，以专门用于存放 Fragment 代码，如图 4.7 所示。

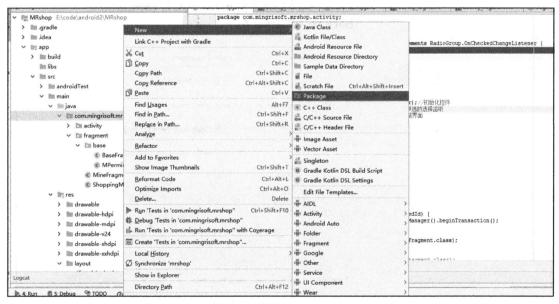

图 4.7　新建 Fragment 包

（3）在 Fragment 包节点上单击鼠标右键，在弹出的快捷菜单中选择新建空白 Fragment，并命名首页碎片为 HomeFragment，如图 4.8 和图 4.9 所示。

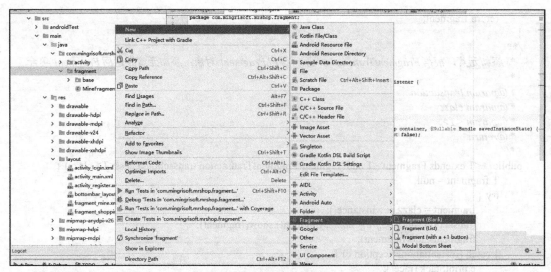

图 4.8 新建 Fragment

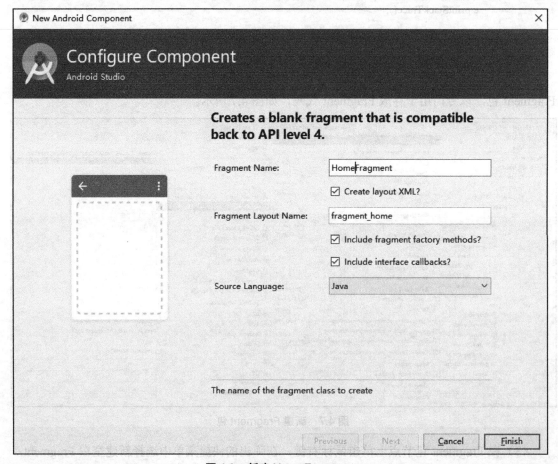

图 4.9 新建 HomeFragment

(4)单击"Finish"按钮,系统自动生成 Fragment 代码和 XML 布局文件。
(5)后续会用到 Fragment 的一些常用方法,为方便开发,此处创建一个公用基础类 BaseFragment。

① 在 Fragment 包内创建 base 包，用于存放公用基础类。
② 在 base 包内，新建基础类 BaseFragment，继承自 Fragment 类。
③ 声明上下文对象和 Activity 对象。
④ 重写 onAttach 方法，当视图被加载到 Activity 中时被调用。
⑤ 重写 onCreateView 等方法，用于初始化控件、设置监听、设置功能等。
具体代码如下：

```java
package com.mingrisoft.mrshop.fragment.base;

import android.APP.Activity;
import android.content.Context;
import android.content.Intent;
import android.os.Bundle;
import android.support.annotation.LayoutRes;
import android.support.annotation.Nullable;
import android.support.v4.APP.Fragment;
import android.view.LayoutInflater;
import android.view.View;
import android.view.ViewGroup;

/**
 * 功能：Fragment 基础类
 */
public abstract class BaseFragment extends Fragment {
    protected Context mContext;         //上下文对象
    protected Activity mActivity;       //Activity 对象

    /**
     * 当视图被加载到 Activity 中时被调用
     * @param context
     */
    @Override
    public void onAttach(Context context) {
        super.onAttach(context);
        this.mContext = context;
        this.mActivity = (Activity) context;
    }

    @Nullable
    @Override
    public View onCreateView(LayoutInflater inflater, @Nullable ViewGroup container, @Nullable Bundle savedInstanceState) {
        View view = inflater.inflate(setLayoutID(),container,false);
        initView(view);//初始化控件
        setListener();//设置监听
        setFunction();//设置功能
        return view;
    }

    /**
     * 设置布局
     * @return
     */
    @LayoutRes
    protected abstract int setLayoutID();
```

```java
/**
 * 跳转界面
 * @param clazz
 * @return
 */
protected final <T extends Activity>Intent activity(Class<T> clazz){
    return new Intent(mActivity,clazz);
}
//设置初始化控件的方法

public void initView(View view) {

}
//设置监听的方法

public void setListener() {

}
//设置功能的方法

public void setFunction() {

}
}
```

（6）在 Fragment 包节点上单击鼠标右键，在弹出的快捷菜单中选择新建 Fragment，并命名为 HomeFragment，继承自 BaseFragment 并实现 View.OnClickListener 监听器接口，重写 setLayoutID 等方法，设置布局页面。具体代码如下：

```java
package com.mingrisoft.mrshop.fragment;

import android.os.Bundle;
import android.support.annotation.Nullable;
import android.view.LayoutInflater;
import android.view.View;
import android.view.ViewGroup;
import android.widget.TextView;

import com.mingrisoft.mrshop.fragment.base.BaseFragment;
import com.mingrisoft.mrshop.mrshop.R;

/**
 * 功能：首页
 */
public class HomeFragment extends BaseFragment implements View.OnClickListener {
    @Override
    protected int setLayoutID() {
        return R.layout.fragment_home;
    }
    @Override
    public View onCreateView(LayoutInflater inflater, @Nullable ViewGroup container, @Nullable Bundle savedInstanceState) {
        View view  = inflater.inflate(setLayoutID(),container,false);
        initView(view);//初始化控件
        setListener();//设置监听
```

```
            return view;
    }
    @Override
    public void initView(View view) {
            super.initView(view);
    }
    /**
     * 设置监听
     */
    @Override
    public void setListener() {

    }

    @Override
    public void onClick(View v) {
            switch (v.getId()) {

            }
    }
}
```

（7）创建一个个人中心页面碎片 MineFragment，用于展示个人中心页面。

（8）返回 MainActivity 类，在选择切换页面的 onCheckedChanged 方法中，选择要显示的 Fragment，如商城首页 HomeFragment 和个人中心页面 MineFragment。

```
/**
 * 选择切换页面
 *
 * @param group
 * @param checkedId
 */
@Override
public void onCheckedChanged(RadioGroup group, int checkedId) {
    FragmentTransaction transaction = getSupportFragmentManager().beginTransaction();
    switch (checkedId) {
        case R.id.mr_shoppingmall://首页
            Toast.makeText(MainActivity.this.getApplicationContext(), "切换至商城首页",Toast.LENGTH_SHORT).show();
            selectFragmentShow(transaction, HomeFragment.class);
            break;
        case R.id.mr_category://分类
            Toast.makeText(MainActivity.this.getApplicationContext(), "切换至商城分类列表",Toast.LENGTH_SHORT).show();
            break;
        case R.id.mr_shoppingcart://购物车
            Toast.makeText(MainActivity.this.getApplicationContext(), "切换至商城购物车", Toast.LENGTH_SHORT).show();
            break;
        case R.id.mr_mine://我的
            Toast.makeText(MainActivity.this.getApplicationContext(), "切换至商城个人中心", Toast.LENGTH_SHORT).show();
            selectFragmentShow(transaction, MineFragment.class);
            break;
    }
    transaction.commit();
}
```

（9）运行该项目，即可实现如图 4.10 所示的商城底部"首页"和"个人中心"的页面切换。

图 4.10 商城底部切换

4.5.2 个人中心页面—登录页面—注册页面的跳转

在个人中心页面，用户要登录才能看到个人的相关信息，需要给 fragment_mine 帧布局的"登录/注册"TextView 组件在 MineFragment 类里添加一个点击监听事件，其作用是若点击则跳转至登录页面。添加步骤是，首先声明 TextView 控件变量，然后重写 initView 方法，初始化绑定 TextView，重写 setListener 方法，在其中为 TextView 设置点击监听器对象，最后重写 onClick 方法实现点击跳转。具体代码如下：

```
private TextView login;//"登录/注册"文本框
@Override
public void initView(View view) {
    super.initView(view);
    login = (TextView) view.findViewById(R.id.custom_login);
}
/**
 * 设置监听
 */
@Override
public void setListener() {
    login.setOnClickListener(this);
}

@Override
public void onClick(View v) {
    switch (v.getId()) {
        case R.id.custom_login://"登录/注册"文本框
            // 跳转至登录页面
            Intent intent = new Intent(getActivity(), LoginActivity.class);
            startActivity(intent);
            break;
    }
}
```

运行该项目，点击"登录/注册"文本框跳转至登录页面，如图 4.11 所示。

图 4.11　个人中心页面—登录页面跳转

在 LoginActivity 中声明、初始化"马上注册"按钮 TextView 文本组件，在 onClick 方法中添加监听点击事件，并跳转至注册页面，具体代码如下：

```
case R.id.mr_reg:
    Intent intent=new Intent(LoginActivity.this, RegisterActivity.class);
    startActivity(intent);
    break;
```

运行该项目，点击"马上注册"按钮即可完成跳转，如图 4.12 所示。

图 4.12　登录页面—注册页面跳转

4.5.3 登录后跳转至个人中心页面

(1) 登录商城后，可以用显式 Intent 的 startActivityForResult 进行 Activity 页面跳转并传递相关数据，重写 LoginActivity 类的 onClick 方法，添加如下代码：

```java
@Override
public void onClick(View view) {
    user=et_user.getText().toString();
    pwd=et_pwd.getText().toString();
    switch (view.getId()) {
        case R.id.mr_login://"登录/注册"文本框
            if("".equals(user)||user==null||"".equals(pwd)||pwd==null) {
                Toast.makeText(LoginActivity.this.getApplicationContext(), "登录失败，账号或密码不能为空", Toast.LENGTH_LONG).show();
            }else{
                Toast.makeText(LoginActivity.this.getApplicationContext(), "登录成功！账号为"+user+",密码为"+pwd, Toast.LENGTH_SHORT).show();
                //登录成功后跳转至指定页面
                Intent intent=new Intent(LoginActivity.this, MainActivity.class);
                //传递数据
                intent.putExtra("user", user);
                intent.putExtra("pwd", pwd);
                startActivity(intent);
            }
            break;
        case R.id.mr_reg:
            Intent intent=new Intent(LoginActivity.this, RegisterActivity.class);
            startActivity(intent);
            break;
    }
}
```

(2) 在 MainActivity 中接收数据并展示。在 onCreate 方法中使用 getIntent 方法获取传递意图，并使用 getStringExtra 方法获取指定参数的对应值。判断用户是否非空，非空则默认跳转至个人中心页面，表示登录成功。具体代码如下：

```java
@Override
protected void onCreate(Bundle savedInstanceState) {
    super.onCreate(savedInstanceState);
    setContentView(R.layout.activity_main);

    // 根据传递过来的意图提取数据
    Intent intent = getIntent();
    String user = intent.getStringExtra("user"); // 获取用户名
    String pwd = intent.getStringExtra("pwd"); // 获取密码

    if(!"".equals(user)&&user!=null&&!"null".equals(user)){
        radioGroup.check(R.id.mr_mine);//已登录并跳转至个人中心页面
    }else{
        radioGroup.check(R.id.mr_shoppingmall);//默认展示商城页面
    }
}
```

(3) 在 MainActivity 类判断跳转至个人中心 MineFragment 后，在 MineFragment 的

onCreateView 方法中根据传递过来的意图提取数据，判断非空后展示个人信息。具体代码如下：

```java
@Override
public View onCreateView(LayoutInflater inflater, @Nullable ViewGroup container, @Nullable Bundle savedInstanceState) {
    View view = inflater.inflate(setLayoutID(),container,false);
    initView(view);//初始化控件
    setListener();//设置监听

    // 根据传递过来的意图提取数据
    Intent intent = getActivity().getIntent();
    String user = intent.getStringExtra("user"); //获取用户名
    String pwd = intent.getStringExtra("pwd"); //获取密码

    if(!"".equals(user)&&user!=null&&!"null".equals(user)){
        Toast.makeText( getActivity(), "跳转成功！账号为"+user+",密码为"+pwd, Toast.LENGTH_LONG).show();
        login.setText(user);
        loginStutas=true;
    }else{
        login.setText("登录/注册");
    }
    return view;
}
```

运行该项目，输入账号及密码后即可成功跳转至个人中心页面，如图 4.13 所示。

图 4.13　登录页面—个人中心页面跳转

4.6　本章小结

本章主要介绍了 Android 应用的重要组成单元 Activity。首先介绍了如何创建、配置、启动和关闭 Activity。实际上，在应用 Android Studio 创建 Android 项目时，就已经默认创建并配置了一个 Activity，如果只需要一个 Activity，则直接使用即可。本章还介绍了多个 Activity 的使用，主要包括如何在两个 Activity 之间交换数据和如何调用另一个 Activity 并返回结果。本章最后介绍了可以合并多个 Activity 的 Fragment。本章内容在实际开发中经常应用，需要重点学习，从而为以后的开发奠定基础。

4.7 本章习题

Activity 与 Fragment 的区别是什么？

第 5 章　数据存储技术

Android 系统提供了多种数据存储方法。例如，使用 SharedPreferences 进行简单存储、文件存储、SQLite 数据库存储等。在本章中将对这几种常用的数据存储方法进行详细的介绍。

 ## 5.1　SharedPreferences 存储

Android 系统提供了轻量级的数据存储方式——SharedPreferences 存储。它屏蔽了对底层文件的操作，通过为程序开发人员提供简单的编程接口，实现以最简单的方式对数据进行永久保存。这种方式主要对少量的数据进行保存，如对应用程序的配置信息、手机应用的主题、游戏的玩家积分等进行保存。例如，对微信进行通用设置后，可以对相关配置信息进行保存，微信通用设置的页面如图 5.1 所示；对手机微博客户端设置应用主题后就可以对该主题进行保存，设置微博主题的页面如图 5.2 所示。

图 5.1　微信通用设置的页面　　　　图 5.2　设置微博主题的页面

5.1.1 获取 SharedPreferences 对象

SharedPreferences 是 Android 系统中的一个轻量级存储类，它位于 android.content 包中，用于使用键值（key-value）对的方式来存储数据。该类主要用于 Boolean、Float、Int、Long、String 等基本类型。在应用程序结束后，数据仍然会保存。数据以 XML 文件格式保存在 Android 系统下"/data/data/<应用程序包名>/shared_prefs"目录中，该文件称为 SharedPreferences（共享的首选项）文件。

获取 SharedPreferences 对象有以下两种方法。

（1）使用 Context 类中的 getSharedPreferences 方法获取。

如果需要使用多个名称来区分 SharedPreferences 文件，则可以使用 getSharedPreferences 方法获取，该方法的基本语法格式如下：

```
getSharedPreferences(String name, int mode)
```

参数说明如下。

- name：共享文件的名称（不包括扩展名），该文件为 XML 格式，用于指定 SharedPreferences 文件的名称，如果指定的文件不存在则会创建一个。
- mode：用于指定操作模式，主要有两种模式可以选择，MODE_PRIVATE 和 MODE_MULTI_PROCESS。MODE_PRIVATE 仍然是默认的操作模式，与直接传入 0 的效果相同，表示只有当前的应用程序才可以对这个 SharedPreferences 文件进行读写。MODE_MULTI_PROCESS 则一般用于多个进程中对同一个 SharedPreferences 文件进行读写的情况。MODE_WORLD_READABLE 和 MODE_WORLD_ WRITEABLE 这两种模式已在 Android 4.2 版本中被废弃。

SharedPreferences 文件都是存放在/data/data/<package name>/shared_prefs/目录下的。

（2）使用 Activity 类中的 getPreferences 方法获取。

如果 Activity 仅有一个默认的 SharedPreferences 文件，则可以直接使用 getPreferences 方法获取。该方法的语法格式如下：

```
getPreferences (int mode)
```

其中，参数 mode 的取值与 getSharedPreferences 方法的相同。

5.1.2 向 SharedPreferences 文件存储数据

向 SharedPreferences 文件中存储数据的步骤如下。

（1）调用 SharedPreferences 类的 edit 方法获得 SharedPreferences.Editor 对象。

（2）向 SharedPreferences.Editor 对象中添加数据。例如，调用 putBoolean 方法添加布尔型数据，调用 putString 方法添加字符串型数据，调用 putInt 方法添加整数型数据。

（3）使用 commit 方法提交数据，从而完成数据存储操作。

5.1.3 读取 SharedPreferences 文件中存储的数据

从 SharedPreferences 文件中读取数据时，主要使用 SharedPreferences 类的 getXxx 方法，

具体代码如下：

```java
//读取所有数据
abstract Map<String, ?> getAll()
//读取的数据类型为 Boolean
abstract boolean    getBoolean(String key, boolean defValue)
//读取的数据类型为 Float
abstract float      getFloat(String key, float defValue)
//读取的数据类型为 Int
abstract int        getInt(String key, int defValue)
//读取的数据类型为 Long
abstract long       getLong(String key, long defValue)
//读取的数据类型为 String
abstract String getString(String key, String defValue)
//读取的数据类型为 Set<String>
abstract Set<String>    getStringSet(String key, Set<String> defValues)
```

介绍了 SharedPreferences 的读取和写入，下面通过具体代码，举例 SharedPreferences 的使用方法。具体代码如下：

```java
/**
 * 保存用户信息
 */
private void saveUserInfo(){
    SharedPreferences userInfo = getSharedPreferences(PREFS_NAME, MODE_PRIVATE);
    SharedPreferences.Editor editor = userInfo.edit();//获取 Editor
    //得到 Editor 后，写入需要保存的数据
    editor.putString("username", "一只猫的涵养");
    editor.putInt("age", 20);
    editor.commit();//提交修改
    Log.i(TAG, "保存用户信息成功");
}
/**
 * 读取用户信息
 */
private void getUserInfo(){
    SharedPreferences userInfo = getSharedPreferences(PREFS_NAME, MODE_PRIVATE);
    String username = userInfo.getString("username", null);//读取用户名
    int age = userInfo.getInt("age", 0);//读取年龄
    Log.i(TAG, "读取用户信息");
    Log.i(TAG, "username:" + username + "，age:" + age);
}
/**
 * 移除年龄的数据
 */
private void removeUserInfo(){
    SharedPreferences userInfo = getSharedPreferences(PREFS_NAME, MODE_PRIVATE);
    SharedPreferences.Editor editor = userInfo.edit();//获取 Editor
    editor.remove("age");
    editor.commit();
    Log.i(TAG, "移除年龄数据");
}
/**
 * 清空数据
 */
private void clearUserInfo(){
```

```
SharedPreferences userInfo = getSharedPreferences(PREFS_NAME, MODE_PRIVATE);
SharedPreferences.Editor editor = userInfo.edit();// 获取 Editor
editor.clear();
editor.commit();
Log.i(TAG, "清空数据");
}
```

5.2 文件存储

学习过 Java SE 的读者都知道，Java 提供了一套完整的 I/O 流体系，通过这些 I/O 流可以很方便地访问磁盘上的文件内容。在 Android 系统中也同样支持通过这种方式来访问手机存储器上的文件。例如，对游戏中需要使用的资源文件进行下载并存储在手机中的指定位置，如图 5.3 所示；再如，将下载的歌曲存储在手机的指定路径下，如图 5.4 所示。

图 5.3　下载并保存资源文件

图 5.4　已下载的歌曲文件

在 Android 系统中主要提供了以下两种方式用于访问手机存储器上的文件。

（1）内部存储。使用 FileOutputStream 类提供的 openFileOutput 方法和 FileInputStream 类提供的 openFileInput 方法访问设备内部存储器上的文件。

（2）外部存储。使用 Environment 类的 getExternalStorageDirectory 方法对外部存储上的文件进行数据读/写。

本节将对这两种方式进行详细讲解。

5.2.1　内部存储

内部存储位于 Android 系统下的 "/data/data/<包名>/files" 目录中。使用 Java 提供的 I/O 流体系可以很方便地对内部存储的数据进行读/写操作。其中，FileOutputStream 类的 openFileOutput 方法用来打开相应的输出流；而 FileInputStream 类的 openFileInput 方法用来打开相应的输入流。默认情况下，使用 I/O 流保存的文件仅对当前应用程序可见，对于其他应用程序（包括用户）是不可见的（不能访问其中的数据）。

 说明：如果用户卸载了应用程序，则保存数据的文件也会一起被删除。

1. 写入文件

若要实现向内部存储器中写入文件，则首先需要获取文件输出流对象 FileOutputStream，这可以使用 FileOutputStream 类的 openFileOutput 方法来实现，然后再调用 FileOutputStream 对象的 write 方法写入文件内容，最后调用 close 方法关闭文件输出流对象。具体代码如下：

```
String fileName = "data.txt";
    String content = "helloworld";
    FileOutputStream fos;
    try {
            //打开应用程序中对应的输出流
            fos = openFileOutput(fileName, MODE_PRIVATE);
            //将数据存储到指定的文件中
             fos.write(content.getBytes());
            //关闭输出流
             fos.close();
    } catch (Exception e) {
             e.printStackTrace();
    }
```

2. 读取文件

若要实现读取内部存储器中的文件，则首先需要获取文件输入流对象 FileInputStream，这可以使用 FileInputStream 类的 openFileInput 方法来实现，然后再调用 FileInputStream 对象的 read 方法读取文件内容，最后调用 close 方法关闭输入流对象。具体代码如下：

```
String content = "";
    FileInputStream fis;
    try {
            //打开应用程序中对应的输入流
            fis = openFileInput("data.txt");
            //创建缓冲区并获取文件长度
            byte[] buffer = new byte[fis.available()];
            //将文件内容读取到 buffer 缓冲区中
            fis.read(buffer);
            //将读取到的内容转换为指定字符串
            content = new String(buffer);
            //关闭输入流
            fis.close();
    } catch (Exception e) {
             e.printStackTrace();
    }
```

5.2.2 外部存储

外部存储是指将文件存储到一些外部设备上（通常位于 mnt/sdcard 目录下，不同厂商生产的手机的路径可能不同），属于永久性的存储方式。每一个 Android 设备都支持共享的外部存储，以用于保存文件，这也是手机中的存储介质。保存在外部存储空间的文件都是全局可读的，而且在使用 USB 连接手机与计算机后，用户可以修改这些文件。

在 Android 程序中，对外部存储的文件进行操作时，需要使用 Environment 类的 getExternalStorageDirectory 方法，该方法用来获取外部存储器的目录。

⚠️ 说明：为了读/写外部存储上的数据，必须在应用程序的全局配置文件（AndroidManifest.xml）中添加读/写外部存储的权限。配置方法如下：

```xml
<uses-permission
        android:name="android.permission.READ_EXTERNAL_STORAGE"/>
<uses-permission
        android:name="android.permission.WRITE_EXTERNAL_STORAGE"/>
```

向外部存储空间写入和读取文件的方法与内部存储类似，都是采用文件输入/输出流的形式。

（1）写入文件。使用 FileOutputStream 文件输出流对象存入数据，具体代码如下：

```java
String state = Environment.getExternalStorageState(); //1.输入流获取外部设备
//2.判断外部设备是否可用并获取SD 卡目录
  if (state.equals(Environment.MEDIA_MOUNTED)) {
            File SDPath = Environment.getExternalStorageDirectory();
            File file = new File(SDPath, "data.txt");
            String data = "HelloWorld";
            FileOutputStream fos;
            try {
                   //3.将数据存储到SD 卡中
                   fos = new FileOutputStream(file);
                   fos.write(data.getBytes());
                   //4.关闭文件输出流对象
                   fos.close();
            } catch (Exception e) {
          e.printStackTrace();
            }
   }
```

（2）读取文件。使用 FileInputStream 文件输入流对象读取数据。具体代码如下：

```java
String state = Environment.getExternalStorageState();//1.获取外部设备
//2.判断外部设备是否可用并获取SD 卡目录
    if (state.equals(Environment.MEDIA_MOUNTED)) {
            File SDPath = Environment.getExternalStorageDirectory();
            File file = new File(SDPath, "data.txt");
            FileInputStream fis;
            try {
                //3.获取指定文件对应的输入流
                fis = new FileInputStream(file);
                //4.将文件输入流对象fis, 并包装为BufferReader 对象
                BufferedReader br = new BufferedReader(new InputStreamReader(fis));
                //5.读取文件内容
                String data = br.readLine();
                //6.关闭输入流
                fis.close();
         } catch (Exception e) {
       e.printStackTrace();
            }
    }
```

5.3　购物商城 APP 的信息存储

5.3.1　用户注册信息的存储

本章学习了文件存储的两种方式，可以用文件存储常用的内部存储来保存用户注册的信息，以下改进用户注册模块，实现用内部文件进行存储用户有效信息的功能。具体步骤如下。

（1）重写 RegisterActivity 类"确定"按钮的监听事件，判断用户信息输入有效后定义内部文件名称 filename 为"usrinfo.txt"，用于存储用户信息。

（2）定义"splitstr"字符，用于分割信息以便存储。

（3）定义输出流 FileOutputStream，并以 MODE_PRIVATE 方式初始化，表示文件只能被创建它的程序访问。

（4）依次写入用户账号、密码、手机号码、邮箱等信息数据，并用 splitstr 分隔符分隔开。

（5）用 fos.close 方法关闭输出流，并提示用户已存储成功。

（6）注册成功后，跳转至登录页面。

具体代码如下：

```java
btn_qr.setOnClickListener(new View.OnClickListener() {
    @Override
    public void onClick(View view) {
        if(edt_pwd.getText().toString().trim().equals(edt_pwd1.getText().toString().trim())
                && !edt_usr.getText().toString().trim().equals("") ){
            //输入的信息有效，注册成功
            Toast.makeText(RegisterActivity.this,"恭喜注册成功！"+edt_usr.getText().toString(),Toast.LENGTH_SHORT).show();
            //将用户信息写入内部文件
            String filename = "usrinfo.txt";
            String splitstr= ",";
            FileOutputStream fos;
            try{
                //写入文件
                fos = openFileOutput(filename,MODE_PRIVATE);
                fos.write(edt_usr.getText().toString().trim().getBytes());
                fos.write(splitstr.getBytes());
                fos.write(edt_pwd.getText().toString().trim().getBytes());
                fos.write(splitstr.getBytes());
                fos.write(edt_phone.getText().toString().trim().getBytes());
                fos.write(splitstr.getBytes());
                fos.write(edt_email.getText().toString().trim().getBytes());
                fos.close();
                Toast.makeText(RegisterActivity.this,"用户信息注册成功写入内部文件！ ",Toast.LENGTH_SHORT).show();
            }catch (Exception e){
                Toast.makeText(RegisterActivity.this,""+e,Toast.LENGTH_SHORT).show();
            }
            Intent intent=new Intent(RegisterActivity.this, LoginActivity.class);
            startActivity(intent);
        }else {
            Toast.makeText(RegisterActivity.this,"输入有误，请重新输入！",Toast.LENGTH_SHORT).show();
        }
    }
});
```

运行该项目，如图 5.5 所示。

图 5.5　用户信息注册成功写入内部文件

5.3.2　免验证快速登录功能

购物商城 APP 在人们的日常生活中经常会用到，如果每次都需要输入账号密码才能使用，那么用户难免会感到麻烦。本节用在本章学到的 SharedPreferences 存储方式来记住密码，实现免验证快速登录功能。

（1）在 activity_login.xml 页面的密码文本框下部增加 CheckBox 组件，用来表示是否记住密码。checked 为"false"表示不记住，作为默认。具体代码如下：

```xml
<CheckBox
    android:id="@+id/mr_check"
    android:layout_width="wrap_content"
    android:layout_height="wrap_content"
    android:checked="false"
    android:text="记住密码" />
```

运行该项目，效果如图 5.6 所示。

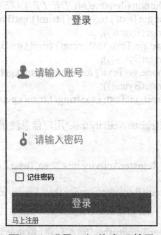

图 5.6　登录—记住密码效果

（2）在后台 LoginActivity 中声明 CheckBox 组件和 SharedPreferences 对象。具体代码如下：

```java
private CheckBox ck_box;          //是否记住密码
private SharedPreferences sp;//声明SharedPreferences对象
```

（3）在 LoginActivity 类的 onCreate 方法中进行初始化，并创建一个名为 UserData 的对象。具体代码如下：

```java
ck_box=(CheckBox)findViewById(R.id.mr_check);
sp = getSharedPreferences("UserData", Context.MODE_PRIVATE);
```

（4）在 LoginActivity 类的 onClick 方法中获取 SharedPreferences 的 edit 对象，通过 edit 对象写入数据，存储当前用户信息并设置记住登录状态（loginStatus 为 "true"），根据 checkBox 的选中状态，更改是否记住密码 isRememberPwd 的状态，最后提交 edit 对象（用 SharedPreferences 存储用户信息，可省略意图传递数据）。具体代码如下：

```java
//获取edit对象
SharedPreferences.Editor edit = sp.edit();

//通过edit对象写入数据，存储当前用户信息
edit.putString("user",user);
edit.putString("pwd",pwd);
edit.putBoolean("loginStatus",true);//记住登录状态

if(ck_box.isChecked()){
    //记住密码选中
    edit.putBoolean("isRememberPwd",true);//是否记住密码
}else{
    //记住密码没有选中
    edit.putBoolean("isRememberPwd",false);//是否记住密码
}

//提交数据存入XML文件中
edit.commit();

Intent intent=new Intent(LoginActivity.this, MainActivity.class);
//用SharedPreferences存储用户信息，可省略意图传递数据
intent.putExtra("user", user);
intent.putExtra("pwd", pwd);
//startActivity(intent);这种启动方式并不能返回结果
startActivityForResult(intent, 1);
```

（5）在登录成功后跳转至 MainActivity 类，在 MainActivity 类中声明 SharedPreferences 对象，并在 onCreate 方法里获取用户信息缓存，在 if 语句中增加判断上次退出应用程序前用户是否为登记状态及是否记住密码，如果满足条件则跳转至个人中心页面。具体代码如下：

```java
private SharedPreferences sp;// 声明SharedPreferences对象
//获取用户信息缓存
sp = this.getSharedPreferences("UserData", Context.MODE_PRIVATE);

if(!"".equals(user)&&user!=null&&!"null".equals(user)){
    radioGroup.check(R.id.mr_mine);//登录成功展示个人中心
}else if(sp.getBoolean("loginStatus",false) && sp.getBoolean("isRememberPwd",false)){
    //判断上次是否为登录状态，以及是否记住密码，如果是则默认跳转至个人中心页面
    radioGroup.check(R.id.mr_mine);
} else{
    radioGroup.check(R.id.mr_shoppingmall);//默认展示商城页面
}
```

（6）跳转至个人中心 MineFragment，在 MineFragment 类中声明登录状态 loginStatus 和 SharedPreferences 对象。具体代码如下：

```java
private Boolean loginStatus;//登录状态
private SharedPreferences sp;//声明 SharedPreferences 对象
```

（7）在 MineFragment 的 initView 方法中初始化 SharedPreferences 对象。具体代码如下：

```java
public void initView(View view) {
    super.initView(view);
    login = (TextView) view.findViewById(R.id.custom_login);
    sp = getActivity().getSharedPreferences("UserData", Context.MODE_PRIVATE);
}
```

（8）在 MineFragment 的 onCreateView 方法里获取用户信息缓存，在 if 语句中增加判断上次退出应用前用户是否为登录状态及是否记住密码，如果满足条件则显示用户信息，并记住当前登录状态（loginStatus 为"true"）。具体代码如下：

```java
if(!"".equals(user)&&user!=null&&!"null".equals(user)){
    Toast.makeText( getActivity(), "跳转个人中心成功！账号为"+user+",密码为"+pwd, Toast.LENGTH_LONG).show();
    login.setText(user);
    loginStatus=true;
}else if(sp.getBoolean("loginStatus",false) && sp.getBoolean("isRememberPwd",false)){
    //上次是否登录及是否记住密码，如果是则加载个人信息
    user=sp.getString("user",null);
    pwd=sp.getString("pwd",null);
    login.setText(user); //展示个人信息
    loginStatus=true; //设置登录状态
}else{
    login.setText("登录/注册");
    loginStatus=false;
}
```

（9）在 MineFragment 类的 onCreateView 方法初始化时，重写并调用 setListener 方法，给"登录/注册"文本框设置单击事件监听，点击"登录/注册"文本框，如果当前用户为非登录状态则跳转至登录页面，如果当前用户为登录状态则弹出对话框询问是否注销当前用户。具体代码如下：

```java
/**
 * 设置监听
 */
@Override
public void setListener() {
    login.setOnClickListener(this);
}

@Override
public void onClick(View v) {
    switch (v.getId()) {
        case R.id.custom_login://登录/注册文本框
            Toast.makeText(getActivity(), "loginStatus:"+loginStatus, Toast.LENGTH_SHORT).show();
            if(loginStatus==true){
                //当前为登录状态时，弹出对话框询问用户是否注销当前用户
                showNormalDialog();
            }else {
```

```
            //当前为非登录状态时，跳转至登录页面
            Intent intent = new Intent(getActivity(), LoginActivity.class);
            startActivity(intent);
        }
        break;
    }
}
```

（10）调用 showNormalDialog 方法，弹出对话框询问用户是否注销当前用户。具体代码如下：

```
/* @setIcon   设置对话框图标
 * @setTitle  设置对话框标题
 * @setMessage 设置对话框消息提示
 */
private void showNormalDialog(){

    final AlertDialog.Builder normalDialog =
            new AlertDialog.Builder(getActivity());

    normalDialog.setTitle("提示");
    normalDialog.setMessage("是否注销当前用户?");
    normalDialog.setPositiveButton("确定",
            new DialogInterface.OnClickListener() {
                //确定按钮回调事件
                @Override
                public void onClick(DialogInterface dialog, int which) {
                    //设置 login 文本内容
                    login.setText("登录/注册");
                    //重置 SharedPreferences 对象、登录状态
                    SharedPreferences.Editor edit = sp.edit();
                    //edit.clear();
                    edit.putBoolean("loginStatus",false);
                    edit.commit();
                    loginStatus=false;
                }
            });
    normalDialog.setNegativeButton("关闭",
            new DialogInterface.OnClickListener() {
                @Override
                public void onClick(DialogInterface dialog, int which) {
                    //关闭按钮回调事件
                }
            });
    //显示
    normalDialog.show();
}
```

（11）再次登录时，在 LoginActivity 类中声明 SharedPreferences 对象，并在 onCreate 方法中获取 SharedPreferences 对象里保存的账号和密码，判断后传递数据并成功跳转至商城首页。具体代码如下：

```
sp = getSharedPreferences("UserData", Context.MODE_PRIVATE);

String user=sp.getString("user","null");//获取当前登录的用户账号
String pwd=sp.getString("pwd","null"); //获取当前登录的密码
if(!"null".equals(user)&& sp.getBoolean("isRememberPwd",false)){
    //默认显示上次的用户
    et_user.setText(user);
```

```
        et_pwd.setText(pwd);
        ck_box.setChecked(true);
    }else{
        //清除所有缓存
        SharedPreferences.Editor edit = sp.edit();
        edit.clear();
        edit.commit();
    }
```

（12）运行该项目，第一次登录后关闭程序重新打开，仍然可以看到上次保存的账号信息，不需要再次验证身份，效果如图 5.7 所示。

图 5.7　记住密码登录效果

5.3.3　退出清除 SharedPreferences

登录成功后，个人中心页面的顶部 textView 会显示当前登录用户的信息。重写 MineFragment 的 login 文本框单击监听事件 onClick 方法，点击判断当前是否为登录状态，如果是，则弹出对话框询问是否注销当前用户，继续选择确定则清除 SharedPreferences 对象，并更改登录状态。具体代码如下：

```
@Override
public void onClick(View v) {
    switch (v.getId()) {
        case R.id.custom_login://登录/注册文本框
            Toast.makeText(getActivity(), "loginStatus:"+loginStatus, Toast.LENGTH_SHORT).show();
            if(loginStatus==true){
                //当前为登录状态时，弹出对话框询问用户是否注销当前用户
                showNormalDialog();
            }else {
                //当前为非登录状态时，跳转至登录页面
                Intent intent = new Intent(getActivity(), LoginActivity.class);
                startActivity(intent);
            }

            break;
    }
}
```

```
/* @setIcon  设置对话框图标
 * @setTitle  设置对话框标题
 * @setMessage  设置对话框消息提示
 * setXXX 方法返回 Dialog 对象,用于设置对话框参数
 */
private void showNormalDialog(){

    final AlertDialog.Builder normalDialog =
            new AlertDialog.Builder(getActivity());

    normalDialog.setTitle("提示");
    normalDialog.setMessage("是否注销当前用户?");
    normalDialog.setPositiveButton("确定",
            new DialogInterface.OnClickListener() {
                //确定按钮回调事件
                @Override
                public void onClick(DialogInterface dialog, int which) {
                    //设置 login 文本内容
                    login.setText("登录/注册");
                    //重置 SharedPreferences 对象、登录状态
                    SharedPreferences.Editor edit = sp.edit();
                    //edit.clear();
                    edit.putBoolean("loginStatus",false);
                    edit.commit();
                    loginStatus=false;
                }
            });
    normalDialog.setNegativeButton("关闭",
            new DialogInterface.OnClickListener() {
                @Override
                public void onClick(DialogInterface dialog, int which) {
                    //关闭按钮回调事件
                }
            });
    //显示
    normalDialog.show();
}
```

5.4 本章小结

本章介绍了 SharedPreferences 对象及其使用方法,讲解了使用内部存储的方式进行文件存储。本章内容在实际开发中经常应用,需要重点学习,以便为以后的开发奠定基础。

5.5 本章习题

简述 SharedPreferences 对象及其使用方法。

第 6 章 数据库编程

6.1 SQLite 数据库简介

Android 系统集成了一个轻量级的关系数据库——SQLite。它不像 Oracle、MySQL 和 SQL Server 等那样专业，但是因为它占用资源少、运行效率高、安全可靠、可移植性强，并且提供零配置运行模式，所以适用于在资源有限的设备（如手机和平板电脑等）上进行数据存取。

在开发手机应用程序时，一般会通过代码来动态地创建数据库，即在程序运行时，首先尝试打开数据库，如果数据库不存在，则自动创建数据库，然后再打开该数据库。下面介绍如何通过代码来创建及操作数据库。

6.2 创建数据库

Android 系统提供了两种创建 SQLite 数据库的方法，下面分别进行介绍。

1. 使用 openOrCreateDatabase 方法创建数据库

SQLiteDatabase 类是 Android 系统中专门用来描述 SQLite 数据库的类。SQLiteDatabase 对象表示一个数据库，应用程序只要获取了代表数据库的 SQLiteDatabase 对象，就可以通过 SQLite- Database 对象来创建数据库。对 SQLiteDatabase 对象调用 openOrCreateDatabase 方法可以打开或创建一个数据库，语法格式如下：

```
static SQLiteDatabase openOrCreateDatabase(String path, SQLiteDatabase.CursorFactory factory)
```

- ◆ path：用于指定数据库文件。
- ◆ factory：用于实例化一个游标。

⚠ 说明：游标提供了一种从表中检索数据并进行操作的灵活手段，通过游标可以一次处理查询结果集中的一行，并可以对该行数据执行特定操作。

例如，使用 openOrCreateDatabase 方法创建一个名为 user.db 的数据库的代码如下：

```
SQLiteDatabase db = SQLiteDatabase. OpenOrCreateDatabase("user.db",null);
```

2. 通过 SQLiteOpenHelper 类创建数据库

在 Android 系统中提供了一个数据库辅助类 SQLiteOpenHelper。通过在该类的构造器中调用 Context 中的方法，可以创建并打开一个指定名称的数据库。在应用这个类时，需要编写继承自 SQLitcOpenHelper 类的子类，并且重写 onCreate 方法和 onUpgrade 方法。

6.3 SQLite 数据库的操作

最常用的数据库的操作包括添加、删除、更新和查询。对于这些操作，程序开发人员完全可以通过执行 SOL 语句来完成。但是这里推荐使用 Android 系统提供的专用类和方法来实现。SQLiteDatabase 类提供了 insert、update、delete 和 query 方法，这些方法封装了执行添加、更新、删除和查询操作的 SQL 命令，所以可以使用这些方法来完成对应的操作，而不用编写 SQL 语句。

（1）添加操作。

SQLiteDatabase 类提供了 insert 方法用于向表中插入数据。insert 方法的基本语法格式如下：

```
public long insert(String table,String nullColumnHack,ContentValues values)
INSERT INTO TABLE_NAME [(column1, column2, column3,...columnN)]
VALUES (value1, value2, value3,...valueN);
```

- ◆ table：用于指定表名。
- ◆ nullColumnHack：可选参数，用于当 values 参数为空时，指定哪个字段设置为 null，如果 values 不为空，则该参数值可以设置为 null。
- ◆ values：用于指定具体的字段值。它相当于 Map 集合，也是通过键值对的形式存储值的。

（2）更新操作。

SQLiteDatabase 类提供了 update 方法用于更新表中的数据。update 方法的基本语法格式如下：

```
public int update(String table,ContentValues values,String whereClause,String[] whereArgs)
UPDATE table_name
SET column1 = value1, column2 = value2,...columnN=valueN
[ WHERE    CONDITION ];
```

- ◆ table：用于指定表名。
- ◆ values：用于指定要更新的字段及对应的字段值。它相当于 Map 集合，也是通过键值对的形式存储值的。
- ◆ whereClause：用于指定条件语句，可以使用占位符（？）。
- ◆ whereArgs：当条件表达式中包含占位符（？）时，该参数用于指定各占位参数的值。如果不包括占位符，则该参数值可以设置为 null。

（3）删除操作。

SQLiteDatabase 类提供了 delete 方法用于从表中删除数据。delete 方法的基本语法格式如下：

```
public int delete(String table, String whereClause,String[] whereArgs)
DELETE FROM table_name
WHERE [conditions...];
```

- table:用于指定表名。
- whereClause:用于指定条件语句,可以使用占位符(?)。
- whereArgs:当条件表达式中包含占位符(?)时,该参数用于指定各占位参数的值。如果不包含占位符,则该参数值可以设置为 null。

(4)查询操作。

SQLiteDatabase 类提供了 query 方法用于查询表中的数据。query 方法的基本语法格式如下:

```
public Cursor query(boolean distinct,String table,String[] columns,String selection,String[] selectionArgs,String groupBy,String having,String orderBy,String limit)
```

- table:用于指定表名。
- columns:用于指定要查询的列。若为空,则返回所有列。
- selection:用于指定 where 子句,即指定查询条件,可以使用占位符(?)。
- selectionArgs:where 子句对应的条件值,当条件表达式中包含占位符(?)时,该参数用于指定各占位参数的值。如果不包含占位符,则该参数值可以设置为 null。
- groupBy:用于指定分组方式。
- having:用于指定 having 条件。
- orderBy:用于指定排序方式,若为空则表示采用默认排序方式。
- limit:用于限制返回的记录条数,若为空则表示不限制。

query 方法的返回值为 Cursor 对象。该对象中保存着查询结果,但是这个结果并不是数据集合的完整复制,而是数据集的指针。通过它提供的多种移动方式,可以获取数据集合中的数据。Cursor 类提供的常用方法如表 6.1 所示。

表 6.1 Cursor 类提供的常用方法

方 法	说 明
moveToFirst()	将指针移动到第一条记录上
moveToNext()	将指针移动到下一条记录上
moveToPrevious()	将指针移动到上一条记录上
getCount()	获取集合的记录数量
getColumnIndexOrThrow()	返回指定字段名称的序号,如果字段不存在,则产生异常
getColumnName()	返回指定序号的字段名称

 ## 6.4 数据信息显示控件

Android 系统常用列表信息显示控件有 ListView、RecyclerView。

6.4.1 ListView 介绍

ListView 有两个重要职责:

（1）将数据填充到布局；
（2）处理用户的选择点击等操作。

ListView 需要三个元素：

（1）ListVeiw 布局：用来展示列表的 View；
（2）适配器：用来把数据映射到 ListView 上的中介；
（3）数据源：具体的将被映射的字符串、图片或基本组件。

适配器是一个连接数据和 AdapterView 的桥梁，通过它能有效地实现数据与 AdapterView 的分离设置，使 AdapterView 与数据的绑定更加简便，修改更加容易。将数据源的数据适配到 ListView 中的常用适配器有 ArrayAdapter、SimpleAdapter 和 SimpleCursorAdapter。

- ArrayAdapter 最为简单，但只能展示一行字。
- SimpleAdapter 有最好的扩充性，可以自定义各种各样的布局，除文本外，还可以存放 ImageView（图像）、Button（按钮）、CheckBox（复选框）等。
- SimpleCursorAdapter 可以认为是 SimpleAdapter 对数据库的简单结合，可以方便地把数据库的内容以列表的形式展示出来。

6.4.2 RecyclerView 介绍

RecyclerView 集成自 ViewGroup 。RecyclerView 是 Android-support-V7 版本中新增加的一个组件，官方对于它的介绍是 RecyclerView 是 ListView 的升级版本，其性能更加先进和灵活。

ListView 只能实现垂直滚动列表，但 RecyclerView 还可以实现水平、多列、跨列等复杂的滚动列表。在使用 RecyclerView 时必须指定一个适配器 Adapter 和一个布局管理器 LayoutManager。适配器继承自 RecyclerView.Adapter 类，具体实现类似 ListView 的适配器，取决于数据信息及展示的 UI。布局管理器用于确定 RecyclerView 中 Item 的展示方式及决定何时复用已经不可见的 Item，避免重复创建及执行高成本的 findViewById 方法。

6.5 购物商城 APP 的数据库编程

购物商城 APP 的数据库编程步骤如下。

- 在包节点上单击鼠标右键，在弹出的快捷菜单中选择新建 entity 包，用于存放购物商城 APP 用到的各种自定义类。
- 在包节点上单击鼠标右键，在弹出的快捷菜单中选择新建 db 包，用于存放数据库操作类。
- 在包节点上单击鼠标右键，在弹出的快捷菜单中选择新建 utils 包，用于存放常用工具类。

新建包的步骤如图 6.1 所示。

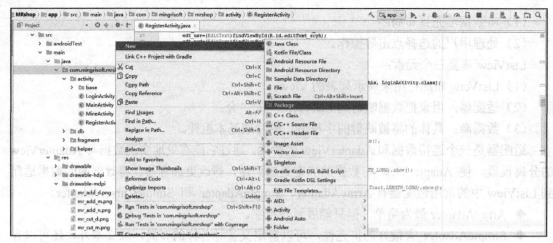

图 6.1 新建包的步骤

相应目录完成后，项目目录结构如图 6.2 所示。

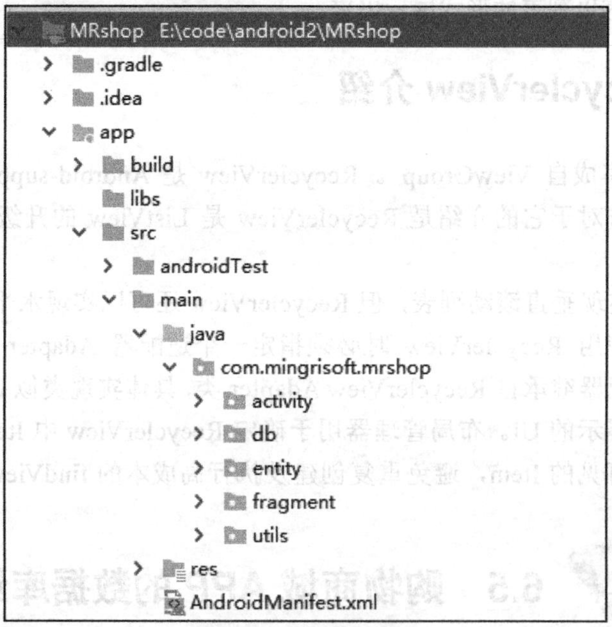

图 6.2 项目目录结构

6.5.1 购物商城 APP 的数据库设计

在前面学习了通过 SQLiteOpenHelper 类创建数据库，在 db 包节点上单击鼠标右键，在弹出的快捷菜单中选择新建 DBHelper 类继承自 SQLiteOpenHelper，并重写 onCreate 方法和 onUpgrade 方法用于创建和更新数据库。具体代码如下：

```
import android.content.Context;
import android.database.sqlite.SQLiteDatabase;
import android.database.sqlite.SQLiteOpenHelper;
import android.util.Log;
```

```java
/**
 * 功能： 创建数据库
 */
public class DBHelper extends SQLiteOpenHelper{
    private static volatile DBHelper dbHelper;
    private static final String DB_NAME = "Shop_db";//数据库的名称
    private DBHelper(Context context) {
        super(context, DB_NAME, null, 1);
    }


    /**
     * 创建数据库
     * @param db
     */
    @Override
    public void onCreate(SQLiteDatabase db) {


    }
    /**
     * 更新数据库
     * @param db
     * @param oldVersion
     * @param newVersion
     */
    @Override
    public void onUpgrade(SQLiteDatabase db, int oldVersion, int newVersion) {


    }
}
```

在 db 包节点上单击鼠标右键，在弹出的快捷菜单中选择新建 DaoUtils 数据操作帮助类，主要有插入数据、更新指定表中的数据、删除表中符合条件的数据、查询数据（需手动关闭数据库连接）、全查询（需手动关闭数据库连接）、关闭数据库连接等常用方法。具体代码如下：

```java
import android.content.ContentValues;
import android.content.Context;
import android.database.Cursor;
import android.database.sqlite.SQLiteDatabase;


import java.lang.reflect.Field;
import java.util.ArrayList;
import java.util.List;

/**
 * 功能： 数据操作帮助类
 */
public class DaoUtils {
    public static final int OPERATION_FAILURE = 0;//操作失败
    public static final int OPERATION_SUCCESS = 1;//操作成功
```

```java
private DBHelper dbHelper;
private SQLiteDatabase database;
public DaoUtils(Context context) {
    dbHelper = DBHelper.getInstance(context);//获取数据库操作类
}


/**
 * 插入数据
 * @param tableName 表的名称
 * @param values 插入的数据（键值对结构）
 * @return 返回操作结果 成功：OPERATION_FAILURE 失败：OPERATION_SUCCESS
 */
public int insert(String tableName, ContentValues values){
    database = dbHelper.getWritableDatabase();
    long result = database.insert(tableName,null,values);
    database.close();
    return result == -1 ? OPERATION_FAILURE : OPERATION_SUCCESS;
}

/**
 * 更新指定表中的数据
 * @param tableName 更改的表名
 * @param values 要更改的数据
 * @param whereClause 筛选条件
 * @param whereArgs 条件
 * @return
 */
public int update(String tableName, ContentValues values, String whereClause, String[] whereArgs){
    database = dbHelper.getWritableDatabase();

    int result = database.update(tableName,values,whereClause,whereArgs);
    database.close();
    return result == 0 ? OPERATION_FAILURE : OPERATION_SUCCESS;
}

/**
 * 删除表中符合条件的数据
 * @param tableName 表名
 * @param whereClause 筛选条件
 * @param whereArgs 条件
 * @return
 */
public int delete(String tableName, String whereClause, String[] whereArgs){
    database = dbHelper.getWritableDatabase();
    int result = database.delete(tableName,whereClause,whereArgs);
    database.close();
    return result == 0 ? OPERATION_FAILURE : OPERATION_SUCCESS;
}

public int deleteAll(String tableName){
    database = dbHelper.getWritableDatabase();
    int result = database.delete(tableName,"",null);
    database.close();
    return result == 0 ? OPERATION_FAILURE : OPERATION_SUCCESS;
}
```

```java
/**
 * 查询数据（需手动关闭数据库连接）
 * @param tableName
 * @param columns
 * @param selection
 * @param selectionArgs
 * @param groupBy
 * @param having
 * @param orderBy
 * @return
 */
public Cursor getResultCursor(String tableName, String[] columns, String selection,
                              String[] selectionArgs, String groupBy, String having,
                              String orderBy){
    database = dbHelper.getWritableDatabase();
    return database.query(tableName,columns,selection,selectionArgs,groupBy,having,orderBy);
}
/**
 * 全查询（需手动关闭数据库连接）
 * @param tableName
 * @return
 */
public Cursor getResultCursorAll(String tableName){
    return getResultCursor(tableName,null,null,null,null);
}

/**
 * 条件查询（需手动关闭数据库连接）
 * @param tableName
 * @param columns
 * @param selection
 * @param selectionArgs
 * @param orderBy
 * @return
 */
public Cursor getResultCursor(String tableName, String[] columns, String selection,
                              String[] selectionArgs,String orderBy){
    return getResultCursor(tableName,columns,selection,selectionArgs,null,null,orderBy);
}
/**
 * 关闭数据库连接
 */
public void closedDB(){
    if (database.isOpen())database.close();
}

/**
 * 返回指定类型的数据集合
 * @param tableName
 * @param clazz
 * @param <T>
 * @return
 */
public <T> List<T> getResultListAll(String tableName,Class<T> clazz){
    List<T> list = new ArrayList<>();
    Cursor cursor = getResultCursorAll(tableName);
    try {
        while (cursor.moveToNext()){
            //创建指定类型的对象
```

```java
                    T obj = clazz.newInstance();
                    Class<?> tClass = obj.getClass();
                    int length = cursor.getColumnCount();//获取列数
                    for (int i = 0;i < length;i++){
                        String ColumnName = cursor.getColumnName(i);//获取列名
                        //根据列名给对象的指定属性赋值
                        Field field = tClass.getDeclaredField(ColumnName);//获取指定的属性
                        setValuesToObject(field,obj,cursor,i);
                    }
                    list.add(obj);
            } catch (InstantiationException e) {
                e.printStackTrace();
            } catch (IllegalAccessException e) {
                e.printStackTrace();
            } catch (NoSuchFieldException e) {
                e.printStackTrace();
            }
            cursor.close();
            closedDB();
            return list;
    }

    /**
     * 设置指定类型的属性值
     * @param field 反射的数据
     * @param object 要生成的对象
     * @param cursor 游标
     * @param index 索引
     */
    private void setValuesToObject(Field field,Object object,Cursor cursor,int index) throws IllegalAccessException {
        String type = field.getType().getSimpleName();//类型
        field.setAccessible(true);//解锁私有属性
        switch (type){//根据类型设置属性值
            case "String":
                field.set(object,cursor.getString(index));
                break;
            case "int":
                field.setInt(object,cursor.getInt(index));
                break;
            case "long":
                field.setLong(object,cursor.getLong(index));
                break;
            case "double":
                field.setDouble(object,cursor.getDouble(index));
                break;
            case "float":
                field.setFloat(object,cursor.getFloat(index));
                break;
        }
    }

    public Cursor getResultBySql(String sql,String[] values){
        database = dbHelper.getWritableDatabase();
        Cursor cursor=database.rawQuery(sql,values);
        return cursor;
    }
```

```java
//判断用户名是否存在
public Boolean isExistUser (String[] values,Boolean isLogin){
    int count=0;
    database = dbHelper.getWritableDatabase();
    String sql="select * from t_user where   name=?";
    //登录时要验证密码
    if(isLogin)
        sql+=" and password=?";
    Cursor cursor=getResultBySql(sql,values);
    count=cursor.getCount();
    cursor.close();
    closedDB();
    return count==0?false:true;
}

/**
 * 广告位图片
 * @return
 */
private String[] bannerUrls() {
    // 获取 banner 图片
    String bannerUrls[] = new String[6];
    String url_start = "banner";
    String url_end = ".jpg";
    url_end="";
    for (int i = 0; i < bannerUrls.length; i++) {
        bannerUrls[i] = url_start + (i + 1) + url_end;
    }
    return bannerUrls;
}
```

6.5.1.1 用户表

（1）通过之前开发的注册功能可以知道，用户表需要账号、密码、姓名、性别、手机号码、邮箱、性别等字段。在 DBHelper 类的 onCreate 方法中创建 t_user 表，用于存储用户信息。具体代码如下：

```java
//用户信息表
String userTable = "create table t_user( name varchar(25) primary key, " +
        "password text, nick text , gender text, phone text," +
        "email text, author text);";
Log.i("创建 user 表", userTable);
db.execSQL(userTable);
```

（2）在 entity 包节点上单击鼠标右键，在弹出的快捷菜单中选择新建 User 类，用于映射 user 表内用户信息。具体代码如下：

```java
public class User {
    private String name;//账号
    private String password;//密码
    private String nick;//姓名
    private String gender;//性别
    private String phone;//手机
    private String email;//邮箱
    private String author;//作者
```

```
//以下省略get、set方法
…
}
```

6.5.1.2 购物车表

（1）购物车数据表主要有主键、商品ID、商品标题、商品价格、商品品牌、图片的保存路径、商品数量、商品图片、商家等字段。在DBHelper类的onCreate方法中创建t_cart表，用于存储购物车信息。具体代码如下：

```
//购物车数据数据表，主键、商品ID、商品标题、商品价格、商品品牌、图片的保存路径、商品数量、商品图片、商家
String cartTable = "create table t_cart(_id varchar(10) primary key, " +
        "title text, price real , brand varchar(20), image_url text," +
        "count integer, image text, merchant varchar(20));";
Log.i("创建 cart 表",cartTable);
db.execSQL(cartTable);
```

（2）在entity包节点上单击鼠标右键，在弹出的快捷菜单中选择新建GoodsCart类，用于映射t_cart表购物车信息。具体代码如下：

```
package com.mingrisoft.mrshop.entity;

/**
 * 功能：购物车中数据
 */
public class GoodsCart {
    private String _id;//商品ID
    private String title;//商品标题
    private double price;//商品价格
    private String brand;//商品品牌
    private String image_url;//图片的保存路径
    private int count;//商品数量
    private String image;//商品图片
    private String merchant;//商家
    private CartViewState viewState;//控件状态
    //以下省略get、set方法
    …
    @Override
    public String toString() {
        return "商品ID:【" + _id +"】商品数量:【" + count +"】";
    }
}
```

（3）在entity包节点上单击鼠标右键，在弹出的快捷菜单中选择新建CartViewState类，用于记录购物车中控件的状态。具体代码如下：

```
package com.mingrisoft.mrshop.entity;

/**
 * 功能：购物车中控件的状态
 */
```

```
public class CartViewState {
    private boolean addViewState;//添加数量控件的状态
    private boolean cutViewState;//减少数量控件的状态
    private boolean checkViewState;//选择控件的状态

    public CartViewState() {
    }

    public CartViewState(boolean addViewState, boolean cutViewState, boolean checkViewState) {
        this.addViewState = addViewState;
        this.cutViewState = cutViewState;
        this.checkViewState = checkViewState;
    }

    //以下省略get、set方法
    ...
    @Override
    public String toString() {
        return "checkViewState(控件选中状态): " + checkViewState;
    }
}
```

6.5.1.3 商品表

（1）商品数据表主要有主键、商品ID、商品名称、商品摘要、商品价格、商品图片、推荐分类、商品类型、商品类型名称、商家等字段。在 DBHelper 类的 onCreate 方法中创建 t_commodity 表，用于存储商品信息。具体代码如下：

```
//commodity 商品表
String commodityTable = "create table t_commodity( id varchar(25) primary key, " +
        "title text, description text , price real, imageUrls    text," +
        "classification text, type integer,typeName text,merchant text);";
Log.i("创建 commodity 表", commodityTable);
db.execSQL(commodityTable);
```

（2）在 entityentity 包中新建 commodity 对象：

```
package com.mingrisoft.mrshop.entity;

/**
 *
 * 功能：商品
 *
 */
//public class Commodity extends Bean{
public class Commodity{
    private String id;//商品ID
    private String title;//商品名称
    private String description;//商品摘要
    private double price;//商品价格
    private String imageUrls;//商品图片
    private String classification;//推荐分类
    private int type;//商品类型
```

```
    private String typeName;//商品类型名称
    private String merchant;//商家

    @Override
    public String toString() {
        StringBuilder stringBuilder = new StringBuilder();
        stringBuilder.APPend("商品ID：").APPend(id).APPend("\n")
                .APPend("名称：").APPend(title).APPend("\n")
                .APPend("摘要：").APPend(description).APPend("\n")
                .APPend("价格：").APPend(price).APPend("\n")
                .APPend("商品图片：").APPend(imageUrls).APPend("\n")
                .APPend("推荐分类：").APPend(classification).APPend("\n")
                .APPend("商品类型：").APPend(type).APPend("\n")
                .APPend("商品类型名称：").APPend(typeName).APPend("\n")
                .APPend("商家：").APPend(merchant).APPend("\n");
        return stringBuilder.toString();
    }
}
```

6.5.1.4 商品详情表

（1）商品详情数据表主要有主键、商品ID、商品名称、商品摘要、商品价格、商品原价、商品图片、商家、提示、商品说明、商品品牌、评论等字段。在 DBHelper 类的 onCreate 方法里创建 t_goodDetails 表，用于存储商品详情信息。

（2）在 entity 包节点上单击鼠标右键，在弹出的快捷菜单上选择新建 GoodDetails 类用于映射商品详情。具体代码如下：

```
package com.mingrisoft.mrshop.entity;

//import com.mingrisoft.mrshop.entity.base.Bean;

/**
 * 功能：商品详情
 */
public class GoodDetails{
//public class GoodDetails extends Bean{
    private String id;// 商品ID
    private String title;// 商品名称
    private String description;//商品摘要
    private double nowPrice;//商品价格
    private double oldPrice;//商品原价
    private String imageUrls;//商品图片
    private String merchant;//商家
    private String prompt;//提示
    private String introduction;//商品说明
    private String brand;//商品品牌
    private int comment;// 评论

    @Override
    public String toString() {
        StringBuilder stringBuilder = new StringBuilder();
        stringBuilder.APPend("商品ID：").APPend(id).APPend("\n")
                .APPend("名称：").APPend(title).APPend("\n")
```

```
                    .APPend("摘要：").APPend(description).APPend("\n")
                    .APPend("价格：").APPend(nowPrice).APPend("\n")
                    .APPend("原价：").APPend(oldPrice).APPend("\n")
                    .APPend("商品图片：").APPend(imageUrls).APPend("\n")
                    .APPend("商家：").APPend(merchant).APPend("\n")
                    .APPend("提示：").APPend(prompt).APPend("\n")
                    .APPend("商品的说明：").APPend(introduction).APPend("\n")
                    .APPend("品牌：").APPend(brand).APPend("\n")
                    .APPend("评论：").APPend(comment).APPend("\n");
            return stringBuilder.toString();
    }
}
```

6.5.2 商品分类模块

（1）在 fragment 节点上单击鼠标右键，在弹出的快捷菜单中选择新建 CategoryFragment 碎片，继承自带有标题栏的公用 TitleFragment 类，并实现 AdapterView.OnItemClickListener，OnItem ClickListener 监听。

（2）声明定义相关全局变量。具体代码如下：

```
private ListView typeList;//类型列表
private RecyclerView shopList;//商品列表
private List<String> typeNames;//类型名称集合
private TypeAdapter nameAdapter;//类型适配器
private CommodityAdapter commodityAdapter;//商品列表适配器
private List<Commodity> commodityList;//商品列表数据
private List<ShapeType> selectList;//选择类型集合
Result<List<String>> result=new Result<List<String>>();
Result<List<Commodity>> resultData=new Result<List<Commodity>>();
private DaoUtils daoUtils;//数据库操作类
```

（3）新建帧布局，并命名为 fragment_category，采用线性布局，左侧为 ListView 组件显示商品类型，右侧用 RecyclerView 加载显示对应类型的商品列表，分类模块布局如图 6.3 所示。具体代码如下：

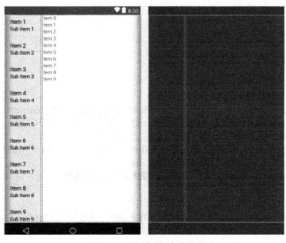

图 6.3 分类模块布局

```xml
<?xml version="1.0" encoding="utf-8"?>
<LinearLayout xmlns:android="http://schemas.android.com/apk/res/android"
    android:orientation="horizontal" android:layout_width="match_parent"
    android:layout_height="match_parent"
    android:background="@color/gray_cancel_btn">
    <ListView
        android:id="@+id/type_list"
        android:layout_width="@dimen/dp_100"
        android:layout_height="match_parent"
        android:divider="@android:color/transparent"
        android:cacheColorHint="@android:color/transparent"
        android:dividerHeight="@dimen/dp_1"
        android:scrollbars="none"/>
    <android.support.v7.widget.RecyclerView
        android:id="@+id/shop_list"
        android:layout_width="match_parent"
        android:layout_height="match_parent"
        android:layout_marginLeft="@dimen/dp_5"
        android:background="@color/gray"
        android:scrollbars="none"/>
</LinearLayout>
```

（4）在 adapter 包节点上单击鼠标右键，在弹出的快捷菜单中选择新建 CommodityAdapter 类，继承自 MyRecyclerViewAdapter 适配器基类。设置适配器布局，绑定数据，更新页面数据。具体代码如下：

```java
/**
 * 功能： 商城列表的适配器
 */

public class CommodityAdapter extends MyRecyclerViewAdapter<Commodity,CommodityAdapter.Commodity ViewHolder> {

    private int layoutRes = R.layout.item_commodity;
    public CommodityAdapter(Context mContext, List<Commodity> mList) {
        super(mContext, mList);
    }
    public CommodityAdapter(Context mContext, List<Commodity> mList, boolean isChange) {
        super(mContext, mList);
        if (isChange) layoutRes = R.layout.item_commodity2;
    }
    /**
     * 绑定数据
     * @param holder
     * @param position
     */
    @Override
    protected void onBindView(final CommodityViewHolder holder, int position) {
        Commodity commodity = mList.get(position);//获取指定位置的数据
        holder.title.setText(commodity.getTitle());//设置标题
        holder.price.setText(FormatUtils.getPriceText(commodity.getPrice()));//设置价格
        String imageUrls = commodity.getImageUrls();
        String url[] = imageUrls.split(StaticUtils.THESEPARATOR);

        //加载图片
        DownLoadImageUtils.loadToImage(mContext,url[0],holder.image);
    }

    /**
```

```java
 * 创建 ViewHoler
 * @param parent
 * @param viewType
 * @return
 */
@Override
public CommodityViewHolder onCreateViewHolder(ViewGroup parent, int viewType) {
    View view = mInflater.inflate(layoutRes,parent,false);
    return new CommodityViewHolder(view);
}

/**
 * 复用类
 */
static class CommodityViewHolder extends RecyclerView.ViewHolder{
    TextView title,price; //标题、价格
    ImageView image; //商品图片
    public CommodityViewHolder(View itemView) {
        super(itemView);
        title = (TextView) itemView.findViewById(R.id.item_commodity_title);
        price = (TextView) itemView.findViewById(R.id.item_commodity_price);
        image = (ImageView) itemView.findViewById(R.id.item_commodity_image);
    }
}
```

（5）CategoryFragment 继承自 TitleFragment，重写 setLayoutID 设置布局；重写 initView 方法初始化控件；重写 setListener 方法设置监听；重写 setFunction 方法设置功能。具体代码如下：

```java
/**
 * 设置布局
 */
@Override
protected int setLayoutID() {
//设置界面布局
    return R.layout.fragment_category;
}

/**
 * 初始化控件
 * @param view
 */
@Override
public void initView(View view) {
    //初始化左侧商品类型列表控件
    typeList = (ListView) view.findViewById(R.id.type_list);
    //初始化右侧商品列表控件
    shopList = (RecyclerView) view.findViewById(R.id.shop_list);
    //为右侧商品列表设置适配器
    shopList.setAdapter(commodityAdapter);
}

@Override
public void setListener() {
//为左侧商品类型列表设置点击监听
    typeList.setOnItemClickListener(this);
}
```

```java
/**
 * 设置功能（此处为了方便查询数据，设置默认商品类型）
 */
@Override
public void setFunction() {
    super.setFunction();
    //设置标题栏
    titleView.setTitle("商品分类");
    //初始化左侧商品类型列表数据
      List<String> lstType=new ArrayList<String>();
    lstType.add("运动护具");
    lstType.add("乳制品");
    lstType.add("粮油调料");

    lstType.add("女鞋");
    lstType.add("男鞋");
    lstType.add("女装");
    lstType.add("户外装备");
    //初始化 result 接收对象
    result =new Result<List<String>>() ;
    result.setReason("成功");
    result.setResult(lstType);
    //给 handler 发送处理消息
    handler.obtainMessage( HttpCode.TYPE_NAME,result).sendToTarget();
}
```

（6）本章项目使用 Handler 消息处理机制来更新 UI。通过 Handler 类可以发送和处理 Message 对象到其所在线程的消息队列中。在 CategoryFragment 类中声明并初始化 handler，根据消息内容进行相应的处理。具体代码如下：

```java
private Handler handler = new Handler(new Handler.Callback() {
    @Override
    public boolean handleMessage(Message msg) {
        switch (msg.what){
            //根据消息类型处理事件
            case HttpCode.TYPE_NAME://类型名称
                //当返回结果为"成功"时，更新左侧类型列表名称集合
                if (TextUtils.equals(HttpCode.SUCCESS, result.getReason())) {
                    typeNames = result.getResult();
                    selectList = new ArrayList<>();
                    //遍历初始化数据
                    for (int i = 0;i < typeNames.size();i++){
                        selectList.add(new ShapeType(typeNames.get(i),false));
                    }
                    //初始化适配器
                    nameAdapter = new TypeAdapter(mContext,selectList);
                    typeList.setAdapter(nameAdapter);
                    typeList.post(new Runnable() {
                        @Override
                        public void run() {
                            View view = typeList.getChildAt(0);
                            typeList.performItemClick(view, 0, typeList.getItemIdAtPosition(0));
                        }
                    });
                }
                break;
```

```
                    case HttpCode.SHOP_LIST://类型商品数据集合
                        if (TextUtils.equals(HttpCode.SUCCESS, resultData.getReason())) {
                            initDataFromNet(resultData); //初始化界面
                        }
                        break;
                }
                return false;
            }
    });
```

（7）加载商品类型到相应适配器中，设置列表显示效果，添加点击监听。具体代码如下：

```
/**
 * 加载数据
 * @param resultData
 */
private void initDataFromNet(Result<List<Commodity>> resultData) {
    if (null == commodityList){
        commodityList = resultData.getResult();
        //创建适配器
        commodityAdapter = new CommodityAdapter(mContext, commodityList,true);
        //设置列表显示效果
        shopList.setLayoutManager(new GridLayoutManager(mContext, 2));
        //设置适配器
        shopList.setAdapter(commodityAdapter);
        //点击监听
        commodityAdapter.setOnItemClickListener(this);
    }else {
        commodityList.clear();
        commodityList.addAll(resultData.getResult());
        commodityAdapter.notifyDataSetChanged();
        shopList.post(new Runnable() {
            @Override
            public void run() {
                shopList.scrollToPosition(0);
            }
        });
    }
}
```

（8）重写商品类型列表点击事件，根据商品类型查询数据，发送消息队列，刷新列表。具体代码如下：

```
/**
 * 点击事件
 * @param parent
 * @param view
 * @param position
 * @param id
 */
@Override
public void onItemClick(AdapterView<?> parent, View view, int position, long id) {
    //获取点击位置，设置适配器
    nameAdapter.setSelectItem(position);
    //根据点击的商品类型查询数据
    downLoadDataFromNet(typeNames.get(position));
```

```java
}
/**
 * 根据类型查询数据
 */
private void downLoadDataFromNet(String shopType) {
    //重新声明 resultData 对象
    resultData=new Result<List<Commodity>>();
    //调用 daoUtils 方法查询数据
    resultData=daoUtils.queryResultCommodityByType(shopType);
    //给 handler 发送消息队列,更新页面
    handler.obtainMessage( HttpCode.SHOP_LIST,resultData).sendToTarget();
}
```

(9) 在 DaoUtils 数据库工具类中添加根据商品类型查询数据的方法。具体代码如下:

```java
public    Result<List<Commodity>> queryResultCommodityByType (String typeName) {
    //初始化返回结果对象
    Result<List<Commodity>> result =new Result<List<Commodity>>();
    //设置返回结果编码
    result.setReason("成功");
    //获取数据库操作对象
    database = dbHelper.getWritableDatabase();
    //查询商品 sql
    String sql="select * from t_commodity where    typeName=?";
    //游标查询
    Cursor cursor=getResultBySql(sql,new String[]{typeName});
    List<Commodity> list =new ArrayList<Commodity>();
    //遍历游标,获取商品对象,添加商品集合
    if(cursor != null){
        while(cursor.moveToNext()) {
            //读取数据
            Commodity commodity=new Commodity();
            commodity.setPrice(cursor.getDouble(cursor.getColumnIndex("price")));
            commodity.setTitle(cursor.getString(cursor.getColumnIndex("title")));
            commodity.setClassification(cursor.getString(cursor.getColumnIndex("classification")));
            commodity.setId(cursor.getString(cursor.getColumnIndex("id")));
            commodity.setDescription(cursor.getString(cursor.getColumnIndex("description")));
            commodity.setType(cursor.getInt(cursor.getColumnIndex("type")));
            commodity.setImageUrls(cursor.getString(cursor.getColumnIndex("imageUrls")));
            commodity.setTypeName(cursor.getString(cursor.getColumnIndex("typeName")));
            commodity.setMerchant(cursor.getString(cursor.getColumnIndex("merchant")));

            list.add(commodity);
        }
    }
    //关闭游标
    cursor.close();
    //关闭数据库
    closedDB();
    //设置并返回结果集
    result.setResult(list);
    return result;
}
```

(10) 重写商品列表的点击事件,模拟跳转至商品详情页面。具体代码如下:

```java
/**
 * 点击事件
 * @param itemView
```

```
 * @param position
 */
@Override
public void onItemClick(View itemView, int position) {
    Toast.makeText(getActivity().getApplicationContext(), "切换至商品详情", Toast.LENGTH_SHORT).show();
}
```

（11）返回 MainActivity 类，在选择切换页面 onCheckedChanged 方法中选择要显示的 Fragment。具体代码如下：

```
case R.id.mr_category://分类
    Toast.makeText(MainActivity.this.getApplicationContext(), "切换至商城分类列表", Toast.LENGTH_SHORT).show();
    selectFragmentShow(transaction, CategoryFragment.class);
    break;
```

（12）运行该项目，效果如图 6.4 所示。

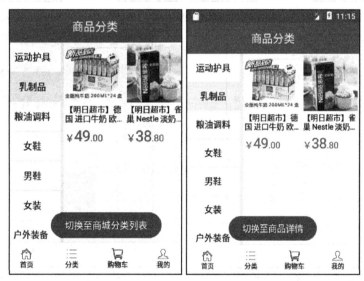

图 6.4 分类模块运行效果图

6.5.3 购物车模块

购物车模块的主要功能和业务逻辑如下。

（1）在购物商城 APP 中，点击某个商品可以进入显示商品的详细信息页面。

（2）在该页面中，点击"加入购物车"按钮将显示确认添加到购物车的窗口。

（3）在该窗口中可以修改商品的购买数量，点击"确认"按钮，即可将该商品添加到购物车中，并关闭窗口，返回商品详细信息页面。

（4）在商品详细信息页面，点击"购物车"按钮，即可显示购物车中的商品数量。

（5）在购物车页面点击商品右侧的"编辑"按钮，可以增加或减少购物车中的商品。

（6）选中要购买的商品，点击"去结算"按钮，将弹出扫码支付的窗口，用户可进行扫码支付。

操作的具体效果，如图 6.5 至图 6.9 所示。

图 6.5 商品详情页面

图 6.6 添加购物车窗口页面

图 6.7 成功添加购物车页面

图 6.8 购物车页面

图 6.9 购物车支付窗口页面

6.6 本章小结

本章首先介绍了什么是 SQLite 数据库，然后介绍了如何创建 SQLite 数据库，最后介绍了在 SQLite 数据库中增加、删除、查询、修改等基础操作。希望读者通过学习本章内容，对 SQLite 数据库的使用有一定的了解。

6.7 本章习题

对 SQLite 数据库进行增加、删除、查询、修改操作的基本语法是什么？

第 7 章 网络编程

智能手机的一个主要功能就是访问互联网。大多数手机 APP 都需要通过互联网执行某种网络通信,因此网络支持对于手机 APP 来说是尤为重要的。本章将对 Android 系统中与网络编程相关的知识进行详细的介绍。

 ## 7.1 通过 HTTP 访问网络

在 Android 系统中可以使用 HTTP 访问网络。例如,在应用商店下载软件时,或者刷新朋友圈时,都需要通过 HTTP 访问网络。

HTTP 的工作原理大致可以理解为客户端向服务器发出一条 HTTP 请求,服务器收到请求后返回一些数据给客户端,客户端对收到的数据进行解析。

在 Android 6.0 版本以前,Android 系统发送 HTTP 请求主要有两种方式:HttpURLConnection 和 HttpClient。其中,HttpClient 存在过多的 API 且难以扩展,于是在 Android 6.0 系统中,HttpClient 被完全移除,如需使用,则需导入相应文件。本章主要介绍 HttpURLConnection 的基本使用方法。

HttpURLConnection 类位于 java.net 包中,用于发送 HTTP 请求和获取 HTTP 响应。由于该类是抽象类,不能直接实例化对象,因此需要使用 URL 的 openConnection 方法来获得。例如,要创建一个 http://www.baidu.com 网站对应的 HttpURLConnection 对象,可以使用下面的代码:

```
Url url= new URL("http://www.baidu.com/");
HttpURLConnection urlConnection=(HttpURLConnection) url.openConnection();
```

HttpURLConnection 是 URLConnection 的一个子类,它在 URLConnection 的基础上提供如表 7.1 所示的常用方法,以方便发送和响应 HTTP 请求。创建 HttpURLConnection 对象后,就可以使用该对象发送 HTTP 请求了。

表 7.1 HttpURLConnection 常用的方法

方 法	描 述
Int getResponseCode()	获取服务器的响应代码
String getResponseMessage()	获取服务器的响应消息
String getRequestMethod()	获取发送请求的方法
void setRequestMethod(String method)	设置发送请求的方法

7.1.1 发送 GET 请求

使用 HttpURLConnection 对象发送请求时,默认发送 GET 请求。因此,发送 GET 请求比较简单,只需要在指定连接地址时,将要传递的参数通过 "?参数名=参数值" 的形式进行传递(多个参数之间可以使用英文半角的&符号分隔),然后获取输入流中的数据,并关闭连接即可。

7.1.2 发送 POST 请求

在 Android 系统中使用 HttpURLConnection 类发送请求时,如果要发送 POST 请求,则需要通过 setRequestMethod 方法进行指定。例如,创建一个 HTTP 连接,并为该连接指定请求的发送方式为 POST,可以使用下面的代码:

```
Url url= new URL("http://www.baidu.com/");
HttpURLConnection urlConn=(HttpURLConnection) url.openConnection();
urlConn.setRequestMethod("POST");
```

发送 POST 请求要比发送 GET 请求复杂一些,它需要通过 HttpURLConnection 类及其父类 URLConnection 提供的方法设置相关内容。发送 POST 请求时常用的方法如表 7.2 所示。

表 7.2 发送 POST 请求时常用的方法

方法	描述
setDoOutput(true); //需要输出	设置是否向连接中写入数据,true 表示写入数据,否则表示不写入数据
setDoInput(true); //需要输入	设置是否从连接中读取数据,true 表示读取数据,否则表示不读取数据
setUseCaches(false); //不允许缓存	设置是否需要缓存数据,true 表示缓存数据,否则表示禁用缓存
SetInstanceFollowRedirects(Boolean follwRedirects)	设置是否应该自动执行 HTTP 重定向,true 表示自动执行,否则不自动执行
setRequestProperty	设置一般请求属性,如果需要设置内容类型为表单数据,则代码为 setRequest Property("Content-Type", "Application/x-www-form-urlencoded");

7.2 解析 JSON 格式数据

7.2.1 JSON 简介

JSON(JavaScript Object Notation)是一种轻量级的数据交换格式,用于数据的标记、存储和传输,其语法简洁,不仅易于阅读和编写,而且也易于机器的解析和生成。JSON 具有以下特点:

- ◆ 轻量级的文本数据交换格式;
- ◆ 独立于语言和平台;
- ◆ 具有自我描述性;
- ◆ 读/写速度快,解析简单。

如下列代码所示，JSON 通常由名称/值（键值对）、对象（名称/值形式的映射）和数组三种方式组成。JSON 没有变量或其他控制，只用于数据传输。

```
{
    "userInfo": {
        "user": [{
            "name": "zhangsan",
            "year": "5"
        },
        {
            "name": "lisi",
            "year": "4"
        }
        ],
        "database": [{
            "name": "shop",
            "size": "2"
        }]
    }
}
```

◆ "名称 / 值" 无序，一个对象用 { } 包括，名称和值间用：相隔，对象间用，隔开：

```
"name": "zhangsan"
```

◆ 对象：一个 JSON 对象包括多个 "名称/值"，书写在花括号 { } 里：

```
{
    "name": "zhangsan",
    "year": "5"
},
```

◆ 数组数组以 [ ] 包括，数据的对象用逗号隔开：

```
[{
    "name": "zhangsan",
    "year": "5"
},
{
    "name": "lisi",
    "year": "4"
}
]
```

7.2.2 解析 JSON 数据

Android 系统提供了用来解析 JSON 数据的 JSONObject 和 JSONArray 对象。其中，JSONObject 用于解析 JSON 对象；JSONArray 用于解析数组。

例如，需要解析的格式为：

```
{
"student":[
            {"id":1,"name":"小明","sex":"男","age":18,"height":175},
            {"id":2,"name":"小红","sex":"女","age":19,"height":165},
```

```
                    {"id":3,"name":"小强","sex":"男","age":20,"height":185}
            ],
    "school":"第一中学"
}
```

则解析的代码如下所示：

```
String jsonStr="…"//jsonStr 为要解析的JSON字符串
  JSONObject root = new JSONObject(jsonStr.toString());
            //根据键名获取键值信息
            System.out.println("root:"+root.getString("cat"));
            JSONArray array = root.getJSONArray("student");
            for (int i = 0;i < array.length();i++)
            {
                    JSONObject stud = array.getJSONObject(i);
                    System.out.println("------------------");
                    System.out.print("id="+stud.getInt("id")+ ",");
                    System.out.print("name="+stud.getString("name")+ ",");
                    System.out.print("sex="+stud.getString("sex")+ ",");
                    System.out.print("age="+stud.getInt("age")+ ",");
                    System.out.println("height="+stud.getInt("height")+ ",");
            }
```

7.3 网络查询手机号码归属地

本章首先介绍了如何通过 HttpURLConnection 访问网络，在本项目中将实现通过 HttpURLConnection 访问网络来查询手机归属地。

（1）在 fragment_mine.xml 中添加一个 LinearLayout 布局，在 LinearLayout 布局中添加两个 ImageView 组件和一个 TextView 组件，用来显示归属地查询功能入口。具体代码如下：

```xml
<!-- 归属地查询 -->
<LinearLayout style="@style/Mine_Item_Style">
    <ImageView
        android:src="@drawable/mr_mine5"
        style="@style/Mine_Image_Style" />
    <TextView
        android:text="归属地查询"
        android:id="@+id/phone_query"
        style="@style/Mine_Text_Style" />
    <ImageView
        android:src="@drawable/mr_right_to"
        style="@style/Mine_Image_Style" />
</LinearLayout>
```

（2）在 MineFragment 类中声明号码归属地 TextView 组件 phone_query，查询结果集 result 及手机号码 phone。在 initView 方法中初始化组件 phone_query，在 setListener 方法中添加点击监听事件。具体代码如下：

```java
private TextView  phone_query;//号码归属地查询组件
String result = "";   //查询结果
String phone = "";    //查询号码
/**
```

```java
 * 初始化组件
 * @param view
 */
@Override
public void initView(View view) {
    super.initView(view);
    login = (TextView) view.findViewById(R.id.custom_login);
    sp = getActivity().getSharedPreferences("UserData", Context.MODE_PRIVATE);
    phone_query = (TextView) view.findViewById(R.id.phone_query);
}

/**
 * 设置监听
 */
@Override
public void setListener() {
    login.setOnClickListener(this);
    phone_query.setOnClickListener(this);
}
```

（3）在 MineFragment 类中初始化 handle 消息对象，用来提示查询消息。具体代码如下：

```java
private Handler handler = new Handler(new Handler.Callback() {
    @Override
    public boolean handleMessage(Message msg) {
        switch (msg.what){
            case 1://1 请求成功
                try {
                    //根据返回值格式，解析返回数据
                    JSONObject jsonObject = new JSONObject(result);
                    JSONObject jsonObject2 = new JSONObject(jsonObject.getString("response"));
                    JSONObject jsonObject3 = new JSONObject(jsonObject2.getString(phone));
                    //提示手机号码归属地
                    Toast.makeText(getActivity(), phone+": "+jsonObject3.getString("location"), Toast.LENGTH_SHORT).show();

                } catch (JSONException e) {
                    e.printStackTrace();
                    Toast.makeText(getActivity(), "查询失败，请重新输入有效号码！", Toast.LENGTH_SHORT).show();
                }
                break;
            case 0://0 请求失败
                Toast.makeText(getActivity(), "查询失败，请重新输入有效号码！", Toast.LENGTH_SHORT).show();
                break;
        }
        return false;
    }
});
```

（4）在 MineFragment 类中新建 getPhoneAddres 方法，用来请求网络数据。具体代码如下：

```java
/**
 * 请求网络数据，获取手机号码归属地
 *
 * @param
 */
public void getPhoneAddres() {
```

```java
//判断手机号码,若不为空,则请求网络接口
    if(phone!=null && !"".equals(phone)&& !"".equals(phone) &&phone.trim().length()>10 ){
        new Thread(new Runnable() {
            @Override
            public void run() {
                try {
                    //接口来源网络
                    URL url = new URL("http://mobsec-dianhua.baidu.com/dianhua_api/open/location?tel="+phone+"&t="+new Date().getTime());
                    //新建 HttpURLConnection 对象
                    HttpURLConnection connection = (HttpURLConnection)url.openConnection();
                    //设置请求类型为get
                    connection.setRequestMethod("GET");
                    //连接、请求时长
                    connection.setConnectTimeout(5000);
                    connection.setReadTimeout(5000);
                    //设置HTTP 请求头,防止乱码
                    connection.setRequestProperty("Content-Type","Application/x-www-form-urlencoded");
                    connection.setRequestProperty("charset", "UTF-8");
                    connection.setRequestProperty("Accept-Charset", "utf-8");
                    connection.setRequestProperty("contentType", "utf-8");
                    StringBuilder s = new StringBuilder();
                    //判断请求是否成功
                    if (connection.getResponseCode() == 200) {
                        InputStreamReader in = new InputStreamReader(connection.getInputStream());
                        BufferedReader buffer = new BufferedReader(in);
                        String inputLine = null;
                        //接收返回结果集
                        while ((inputLine = buffer.readLine()) != null) {
                            result += inputLine;
                        }
                        //接收完消息,发送handler 消息队列,提示消息
                        handler.obtainMessage( 1,result).sendToTarget();
                    }
                } catch (Exception e) {
                    e.printStackTrace();
                }
            }
        }).start();
    }else{
        handler.obtainMessage( 0,result).sendToTarget();
    }
}
```

(5)在 MineFragment 类中新建 showPhoneDialog 方法,用来展示查询手机号码对话框。点击"确定"按钮后,初始化手机号码并返回结果,同时触发 getPhoneAddres 方法请求网络数据。具体代码如下:

```java
/* @setIcon 设置对话框图标
 * @setTitle 设置对话框标题
 * @setMessage 设置对话框消息提示
 * setXXX 方法返回 Dialog 对象,因此可以设置属性
 */
private void showPhoneDialog() {
    final AlertDialog.Builder normalDialog =
            new AlertDialog.Builder(getActivity());
    final EditText et = new EditText(getActivity());
```

```
        normalDialog.setTitle("提示");
        normalDialog.setView(et);
        normalDialog.setMessage("请输入查询的手机号码?");
        normalDialog.setPositiveButton("确定",
                new DialogInterface.OnClickListener() {
//重写"确定"按钮点击事件,点击"确定"按钮,根据号码请求网络接口,获取手机号码归属地
                    @Override
                    public void onClick(DialogInterface dialog, int which) {
                        phone=et.getText().toString();
                        result = "";
                        getPhoneAddres();
                    }
                });
        normalDialog.setNegativeButton("关闭",
                new DialogInterface.OnClickListener() {
                    @Override
                    public void onClick(DialogInterface dialog, int which) {
                        //...To-do
                        result = "";
                    }
                });
        // 显示
        normalDialog.show();
    }
```

（6）重写 MineFragment 中的 onClick 方法，设置归属地查询组件 phone_query 的点击事件，触发 showPhoneDialog 方法，显示查询号码对话框，进行手机归属地查询，发送消息队列。具体代码如下：

```
@Override
public void onClick(View v) {
    switch (v.getId()) {
        case R.id.custom_login://登录/注册文本框
            Toast.makeText(getActivity(), "loginStatus:" + loginStatus, Toast.LENGTH_SHORT).show();
            if (loginStatus == true) {
                showNormalDialog();
            } else {
                Intent intent = new Intent(getActivity(), LoginActivity.class);
                startActivity(intent);
            }
            break;
        case R.id.phone_query://归属地查询
            showPhoneDialog();
            break;
    }
}
```

（7）在 AndroidManifest.xml 配置中心添加访问网络权限。具体代码如下：

```
<uses-permission android:name="android.permission.INTERNET"/>
```

（8）运行该项目，效果如图 7.1 所示。

图 7.1　手机号码归属地查询效果

 7.4　本章小结

本章首先介绍了如何通过 HTTP 访问网络，主要是使用 java.net 包中的 HttpURLConnection 实现；然后介绍了 JSON 的主要特点、组成方式及如何解析 JSON 格式数据。

 7.5　本章习题

GET 和 POST 请求的区别是什么？

第 8 章 多媒体编程

本章将对 Android 系统中的补间动画、逐帧动画,以及音频、视频等多媒体应用进行详细介绍。

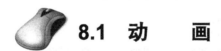 8.1 动　　画

在应用 Android 系统进行项目开发时,特别是在进行游戏开发时,经常需要涉及动画。Android 系统中的动画通常可以分为补间动画和逐帧动画两种。下面将分别介绍如何实现这两种功能。

8.1.1 补间动画

补间动画（Tween Animation）通过对 View 的内容进行一系列的图形变换（包括平移、缩放、旋转、改变透明度）来实现动画效果,动画类型如表 8.1 所示。动画效果可以采用 XML 或编码来实现。

表 8.1　动画类型

动 画 类 型	XML 配置方式	Java 代码实现方式
渐变透明度动画效果	<alpha/>	AlphaAnimation
渐变尺寸缩放动画效果	<scale/>	ScaleAnimation
画面旋转动画效果	<rotate/>	RotateAnimation
画面位置移动动画效果	<translate/>	TranslateAnimation
组合动画效果	<set/>	AnimationSet

（1）alpha 渐变透明度动画效果 XML 布局如下：

```xml
<?xml version="1.0" encoding="utf-8"?>
<alpha xmlns:android="http://schemas.android.com/apk/res/android"
    android:duration="500"
    android:fillAfter="false"
    android:fromAlpha="1.0"
    android:toAlpha="0.0" />
```

alpha 渐变透明度动画属性及描述如表 8.2 所示。

表 8.2 alpha 渐变透明度动画属性及描述

属性	描述
fromAlpha	开始时的透明度
toAlpha	开始时的透明度
fillAfter	结束时的透明度
fromAlpha	动画的持续时间
duration	设置动画结束后保持当前的位置

XML 方式加载方式通过 AnimationUtils.loadAnimation(this,R.anim.anim_alpha)获取 Animation。

```
Animation alphaAnimation = AnimationUtils.loadAnimation(this, R.anim.anim_alpha);
    imageView.startAnimation(alphaAnimation);
```

（2）scale 渐变尺寸缩放动画效果 XML 布局如下：

```xml
<?xml version="1.0" encoding="utf-8"?>
<scale xmlns:android="http://schemas.android.com/apk/res/android"
    android:duration="500"
    android:fromXScale="0.0"
    android:fromYScale="0.0"
    android:interpolator="@android:anim/decelerate_interpolator"
    android:pivotX="50%"
    android:pivotY="50%"
    android:repeatCount="1"
    android:repeatMode="reverse"
    android:startOffset="0"
    android:toXScale="1.5"
    android:toYScale="1.5" />
```

scale 渐变尺寸缩放动画效果如表 8.3 所示。

表 8.3 scale 渐变尺寸缩放动画效果

属性	描述
fromXDelta，fromYDelta	起始时 X、Y 坐标，屏幕右下角的坐标是 X:320,Y:480
toXDelta，toYDelta	动画结束时 X、Y 坐标
interpolator	指定动画插入器
fromXScale,fromYScale	动画开始前 X、Y 的缩放，0.0 为不显示，1.0 为正常大小
toXScale，toYScale	动画最终缩放的倍数，1.0 为正常大小，大于 1.0 为放大
pivotX，pivotY	动画起始位置相对于屏幕的百分比，两个都为 50%表示动画从自身中间开始
startOffset	动画多次执行的间隔时间，如果只执行一次，则执行前会暂停这段时间，单位为毫秒
duration	一次动画效果消耗的时间，单位为毫秒，值越小则动画效果消耗时间越短
repeatCount	动画重复的计数，动画将会执行该值+1 次
repeatMode	动画重复的模式，reverse 为反向，当第偶次执行时，动画的执行方向会相反。restart 为重新执行，方向不变

在动画的每个周期里面做不同的操作，可以借助动画监听器 Animation.AnimationListener。具体代码如下：

```
alphaAnimation.setAnimationListener(new Animation.AnimationListener() {
    @Override
    public void onAnimationStart(Animation animation) {
```

```
        //动画开始时调用
    }
    @Override
    public void onAnimationEnd(Animation animation) {
        //动画结束时调用
    }
    @Override
    public void onAnimationRepeat(Animation animation) {
        //动画重复时调用
    }
});
```

8.1.2 逐帧动画

逐帧动画（Frame-by-Frame Animations）从字面上理解就是一帧挨着一帧地播放图片，就像放电影一样。逐帧动画和补间动画一样，可以通过 XML 实现也可以通过 Java 代码实现。接下来，借助目前项目中的一个开奖的动画来总结如何实现逐帧动画。实现效果如图 8.1 所示。

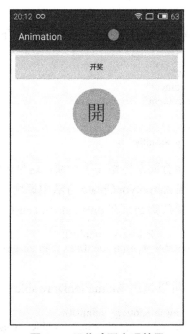

图 8.1 开奖动画实现效果

1. lottery_animlist, xml 文件

在 res/drawable 目录中新建一个文件 lottery_animlist.xml，其内容如下：

```xml
<?xml version="1.0" encoding="utf-8"?>
<animation-list xmlns:android="http://schemas.android.com/apk/res/android"
    android:oneshot="false">
    <item
        android:drawable="@mipmap/lottery_1"
        android:duration="200" />
    <item
        android:drawable="@mipmap/lottery_2"
        android:duration="200" />
```

```xml
    <item
        android:drawable="@mipmap/lottery_3"
        android:duration="200" />
    <item
        android:drawable="@mipmap/lottery_4"
        android:duration="200" />
    <item
        android:drawable="@mipmap/lottery_5"
        android:duration="200" />
    <item
        android:drawable="@mipmap/lottery_6"
        android:duration="200" />
</animation-list>
```

根节点是 animation-list（动画列表），里面由一个或多个 item 节点组成；oneshot 属性表示是否只播放一次，true 表示只播放一次，false 表示一直循环播放；内部用 item 节点声明一个动画帧；android:drawable 指定此帧动画所对应的图片资源；android:druation 代表此帧持续的时间，值为整数，单位为毫秒。

2. 显示动画

用 ImagView 控件作为动画载体来显示动画，具体代码如下：

```xml
<ImageView
    android:id="@+id/animation_iv"
    android:layout_width="wrap_content"
    android:layout_height="wrap_content"
    android:layout_gravity="center"
    android:layout_margin="10dp"
    android:src="@drawable/lottery_animlist" />
```

运行该动画，发现动画没有运行而是停留在第一帧，这是因为 AnimationDrawable 播放动画是依附在 Window 上面的，而在 ActivityonCreate 方法中调用时 Window 还未初始化完毕，所以才会停留在第一帧，要想实现播放则必须在 onWindowFocusChanged 中添加如下代码：

```java
imageView.setImageResource(R.drawable.lottery_animlist);
AnimationDrawable animationDrawable = (AnimationDrawable) imageView.getDrawable();
animationDrawable.start();
```

如果想要停止播放动画，则可以调用 AnimationDrawable 的 stop 方法：

```java
imageView.setImageResource(R.drawable.lottery_animlist);
    AnimationDrawable animationDrawable = (AnimationDrawable) imageView.getDrawable();
    animationDrawable.stop();
```

3. 纯 Java 代码实现方式

纯 Java 代码实现方式如下：

```java
AnimationDrawable anim = new AnimationDrawable();
    for (int i = 1; i <= 6; i++) {
    int id = getResources().getIdentifier("lottery_" + i, "mipmap", getPackageName());
    Drawable drawable = getResources().getDrawable(id);
    anim.addFrame(drawable, 200);
    }
    anim.setOneShot(false);
    imageView.setImageDrawable(anim);
    anim.start();
```

4. AnimationDrawable 常见的 API

AnimationDrawable 常见的 API 如表 8.4 所示。

表 8.4 AnimationDrawable 常见 API

方　　法	描　　述
void start()	开始播放动画
void stop()	停止播放动画
addFrame(Drawable frame, int duration)	添加一帧，并设置该帧显示的持续时间
void setOneShoe(boolean flag)	false 为循环播放，true 为仅播放一次
boolean isRunning()	是否正在播放

逐帧动画相对来讲比较简单，但是在实际开发中使用的频率还是比较高的，希望通过这个小例子能够使读者掌握逐帧动画。由于逐帧动画只能实现比较小的动画效果，因此对于复杂且帧数比较多的动画，则不太建议使用逐帧动画。如果是超级复杂的动画，则建议选择双缓冲绘制 View 来实现。

8.2　音频与视频

Android 系统提供了对常用音频和视频格式的支持，所支持的音频格式有 MP3、3GP、Ogg 和 WAVE 等，支持的视频格式有 3GP 和 MP4 等。利用 Android API 提供的相关方法，可以在 Android 应用程序中实现音频与视频的播放。下面将分别介绍播放音频与视频的不同方法。

8.2.1　使用 MediaPlayer 类播放音频

MediaPlayer 是一个支持音频及视频文件播放的 Android 类，可播放不同来源（本地或网络流媒体）、多种格式（如 WAV、MP3、Ogg Vorbis、MPEG-4 及 3GPP）的多媒体文件。要想利用 MediaPlayer 类实现音频的播放，则首先要对 MediaPlayer 类进行初始化工作，得到 MediaPlayer 对象，再通过 MediaPlayer 对象进行相应的操作。

一般过程是，初始化 MediaPlayer→加载媒体源→准备→开始播放：

```
MediaPlayer mediaPlayer = new MediaPlayer();
mediaPlayer.setDataSource("...");
mediaPlayer.prepare();
mediaPlayer.start();
```

MediaPlayer 类支持多种不同的媒体源，包括本地资源、内部的 URI。例如，从 ContentResolver 获取的 URI、外部 URL（流）。

raw 文件中媒体源：假如 res/raw 文件中包含一个 sound_music.mp3 文件：

```
MediaPlayer mediaPlayer = MediaPlayer.create(this, R.raw.sound_music);
```

assets 文件中媒体源：假如在 assets 中包含一个 sound_music.mp3 文件：

```
try {
    AssetFileDescriptor fd = getAssets().openFd("sound_music.mp3");
    MediaPlayer mediaPlayer = new MediaPlayer();
```

```
        mediaPlayer.setDataSource(fd.getFileDescriptor(), fd.getStartOffset(), fd.getLength());
} catch (IOException e) {
        e.printStackTrace();
}
```

SD 卡中媒体源：假如在 SD 卡中包含一个 sound_music.mp3 文件：

```
try {
        MediaPlayer mediaPlayer = new MediaPlayer();
        String path = "/sdcard/sound_music.mp3";
        mediaPlayer.setDataSource(path);
} catch (IOException e) {
        e.printStackTrace();
}
```

网络资源：假如有一个网络资源 http://ibooker.cc/ibooker/musics/sound_music.mp3：

```
MediaPlayer mediaPlayer = new MediaPlayer();
// 方式一
//     Uri uri = Uri.parse("http://ibooker.cc/ibooker/musics/sound_music.mp3");
//     mediaPlayer.setDataSource(this, uri);

// 方式二
mediaPlayer.setDataSource("http://ibooker.cc/ibooker/musics/sound_music.mp3");
```

MediaPlayer 类常用方法如下：

```
int getCurrentPosition();// 得到当前播放位置（以毫秒为单位）
int getDuration();// 得到文件的时间（以毫秒为单位）
void setLooping(boolean var1);// 设置是否循环播放
boolean isLooping();// 是否循环播放
boolean isPlaying();// 是否正在播放
void pause();// 暂停
void prepare();// 同步准备
void prepareAsync();// 异步准备
void release();// 释放 MediaPlayer 对象
void reset();// 重置 MediaPlayer 对象
void seekTo(int msec);// 指定播放位置（以毫秒为单位）
void setDataSource(String path);// 设置播放资源
void setScreenOnWhilePlaying(boolean screenOn);// 设置播放时一直使屏幕亮
void setWakeMode(Context context, int mode);// 设置唤醒模式
void setVolume(float leftVolume, float rightVolume);// 设置音量，参数分别表示左右声道声音大小，取
值范围为 0~1
void start();// 开始播放
void stop();// 停止播放
```

MediaPlayer 常用事件监听如下。

◆ 播放出错监听：

```
MediaPlayer.setOnErrorListener(new MediaPlayer.OnErrorListener() {
        @Override
        public boolean onError(MediaPlayer mediaPlayer, int i, int i1) {
                return false;
        }
});
```

◆ 播放完成监听：

```
MediaPlayer.setOnCompletionListener(new MediaPlayer.OnCompletionListener() {
    @Override
    public void onCompletion(MediaPlayer mediaPlayer) {
        // todo
    }
});
```

◆ 网络流媒体缓冲监听：

```
MediaPlayer.setOnBufferingUpdateListener(new MediaPlayer.OnBufferingUpdateListener() {
    @Override
    public void onBufferingUpdate(MediaPlayer mediaPlayer, int i) {
        // i 为 0~100
        Log.d("Progress:", "缓存进度" + i + "%");
    }
});
```

◆ 准备 Prepared 完成监听：

```
MediaPlayer.setOnPreparedListener(new MediaPlayer.OnPreparedListener() {
    @Override
    public void onPrepared(MediaPlayer mediaPlayer) {
        // todo
    }
});
```

◆ 进度调整完成 SeekComplete 监听，主要配合 seekTo 方法：

```
MediaPlayer.setOnSeekCompleteListener(new MediaPlayer.OnSeekCompleteListener() {
    @Override
    public void onSeekComplete(MediaPlayer mediaPlayer) {
        // todo
    }
});
```

8.2.2 使用 SoundPool 类播放视频

Android 系统还提供了另外一个播放音频的类——SoundPool（音频池），用于同时播放多个短小的音频，而且占用的资源较少。SoundPool 类主要用于播放一些较短的声音片段，与 MediaPlayer 类相比，SoundPool 类的优势在于 CPU 资源占用量低和反应延迟小。另外，SoundPool 类还支持自行设置声音的品质、音量、播放比率等参数。基本方法如下。

构造器：用于初始化一个 SoundPool 类：

```
SoundPool(int maxStreams, int streamType, int srcQuality)//
```

◆ maxStreams：指定同时可以播放的音频流个数。
◆ streamType：指定声音的类型，简单地讲就是，播放时以哪种声音类型的音量播放。例如，STREAM_ALARM 是警报的声音类型。
◆ srcQuality：音频的质量，设置为 0 代表默认。

加载音频：提供不同的加载方式，可以从 res/raw 中加载，或者通过文件的绝对路径加载，并指定优先级。优先级越高就越优先播放。加载完成后返回一个资源 ID，代表这个音频在 SoundPool 类中的 ID：

　　int load(Context context, int resId, int priority)：

播放音频：需要指定加载时返回的资源 ID 才能播放：

　　int play(int soundID, float leftVolume, float rightVolume, int priority, int loop, float rate)

- soundID：该方法的第一个参数指定播放哪个声音，即 load 后返回的 ID。
- leftVolume、 rightVolume：指定左、右声道的音量。
- priority：指定播放声音的优先级，数值越大优先级就越高。
- loop：指定是否循环，0 为不循环，–1 为循环。
- rate：指定播放的比率，数值可从 0.5 到 2，1 为正常比率。

加载完成的回调：虽然是加载一个很小的音频，但还是需要一点时间。所以，就需要这个回调：

　　onLoadComplete(SoundPool soundPool, int sampleId, int status)

- sampleId 就是音频的 ID，用于标识哪个音频。
- status 是加载完成的状态，0 为成功。

8.2.3　使用 VideoView 组件播放视频

在 Android 系统中提供了 VideoView 组件用于播放视频文件，要想使用 VideoView 组件播放视频则首先要在布局文件中添加该组件，然后在 Activity 中获取该组件，并应用其 setVideoPath 方法或 setVideoURI 方法加载要播放的视频，最后调用 start 方法来播放视频。另外，VideoView 组件还提供了 stop 方法和 pasuse 方法，分别用于通知和暂停视频的播放。基本使用格式如下：

```
VideoView mVv = (VideoView) findViewById(R.id.vv);
//添加播放控制条
mVv.setMediaController(new MediaController(this));
//设置视频源播放 res/raw 中的文件,文件名小写字母,格式包括 3GP、MP4 等，不一定支持 FLV 格式
Uri rawUri = Uri.parse("android.resource://" + getPackageName() + "/" + R.raw.shuai_dan_ge);
mVv.setVideoURI(rawUri);

//播放在线视频
mVideoUri = Uri.parse("http://****/abc.mp4");
mVv.setVideoPath(mVideoUri.toString());
mVv.start();
mVv.requestFocus();
mVv.resume();
mVv.setOnPreparedListener(this);
mVv.setOnErrorListener(this);
mVv.setOnCompletionListener(this);
```

8.3　商品详情页面的背景音乐

本章学习了使用 MediaPlayer 类播放音频，介绍了如何使用 MediaPlayer 类给商品详情页面添加背景音乐。本章代码涉及 Service 服务，因为 Service 是能够在后台长时间运行，并且

不提供用户页面的应用程序组件。当其他应用程序组件启动 Service，并且切换到另一个应用程序时，Service 还可以在后台运行。如果要关闭 Service 和背景音乐，则可以重写 onBackPressed 方法，监听返回按键，如果返回键被按下则关闭服务。

（1）在项目名上单击鼠标右键，在弹出的快捷菜单中选择新建 service 包，用于存放 service。新建 service 包如图 8.2 所示。

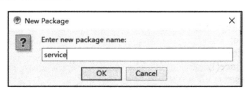

图 8.2　新建 service 包

（2）在 AndroidManifest.xml 文件中的<Application>标签内加入以下语句：

```xml
<service android:name="com.mingrisoft.mrshop.service.MusicServer">
    <intent-filter>
        <action android:name="com.angel.Android.MUSIC"/>
        <category android:name="android.intent.category.default" />
    </intent-filter>
</service>
```

（3）在 service 包节点上单击鼠标右键，在弹出的快捷菜单中选择新建 MusicServer.java 类，继承自 Service。

（4）声明 mediaPlayer，重写 onCreate 方法以初始化媒体播放器；重写 onStartCommand 方法以启动播放；重写 onDestroy 方法以停止播放。

具体代码如下：

```java
import android.APP.Service;
import android.content.Intent;
import android.media.MediaPlayer;
import android.os.IBinder;
import android.widget.Toast;

import java.io.IOException;

public class MusicServer extends Service {

    private MediaPlayer mediaPlayer = null;

    private boolean isReady = false;

    @Override
    public void onCreate() {
        //onCreate 在 Service 的生命周期中只会调用一次
        super.onCreate();

        //初始化媒体播放器
        mediaPlayer = MediaPlayer.create(this, R.raw.shop_music);
        if(mediaPlayer == null){
            return;
        }

        mediaPlayer.stop();
        mediaPlayer.setOnErrorListener(new MediaPlayer.OnErrorListener() {
```

```java
            @Override
            public boolean onError(MediaPlayer mp, int what, int extra) {
                mp.release();
                stopSelf();
                return false;
            }
        });

        try{
            mediaPlayer.prepare();
            isReady = true;
        } catch (IOException e) {
            e.printStackTrace();
            isReady = false;
        }

        if(isReady){
            //将背景音乐设置为循环播放
            mediaPlayer.setLooping(true);
        }
    }

    @Override
    public int onStartCommand(Intent intent, int flags, int startId) {
        //每次调用 Context 的 startService 都会触发 onStartCommand 回调方法
        //所以 onStartCommand 在 Service 的生命周期中可能会被多次调用
        if(isReady && !mediaPlayer.isPlaying()){
            //播放背景音乐
            mediaPlayer.start();
            Toast.makeText(this, "开始播放背景音乐", Toast.LENGTH_LONG).show();
        }
        return START_STICKY;
    }

    @Override
    public IBinder onBind(Intent intent) {
        //该 Service 中不支持 bindService 方法，所以此处直接返回 null
        return null;
    }

    @Override
    public void onDestroy() {
        //当调用 Context 的 stopService 或 Service 内部执行 stopSelf 方法时就会触发 onDestroy 回调方法
        super.onDestroy();
        if(mediaPlayer != null){
            if(mediaPlayer.isPlaying()){
                //停止播放音乐
                mediaPlayer.stop();
            }
            //释放媒体播放器资源
            mediaPlayer.release();
            Toast.makeText(this, "停止播放背景音乐", Toast.LENGTH_LONG).show();
        }
    }
}
```

（5）在商品详情 GoodsDetailsActivity 的 onCreate 方法里启动播放。具体代码如下：

```java
Toast.makeText(GoodsDetailsActivity.this, "开始播放背景音乐", Toast.LENGTH_SHORT).show();
//播放背景音乐
Intent intent = new Intent(GoodsDetailsActivity.this, MusicServer.class);
this.startService(intent);
```

（6）重写 onBackPressed 方法，以便监听返回键是否被按下，从而判断是否应关闭背景音乐。具体代码如下：

```
/**
* 注意:
* super.onBackPressed()会自动调用 finish 方法,关闭当前 Activity
*/
@Override
public void onBackPressed() {
    super.onBackPressed();
    Intent intent2 = new Intent(GoodsDetailsActivity.this, MusicServer.class);
    stopService(intent2);// 关闭服务
    Toast.makeText(GoodsDetailsActivity.this, "按下了 back 键 关闭背景音乐！", Toast.LENGTH_SHORT).show();
}
```

（7）运行该项目，运行效果如图 8.3 所示。

图 8.3 播放背景音乐效果

 8.4 本章小结

本章主要介绍了在 Android 中如何使用动画，以及如何播放音频和视频。在 Android 系统开发过程中，动画与多媒体技术的应用非常广泛，希望读者能够通过加强练习的方式来完全掌握本章的内容。

 8.5 本章习题

mediaplayer 与 soundPool 的区别是什么？

第9章 实现购物商城APP的其他功能

 9.1 用户身份验证与注册

（1）如果注册时需要验证数据库内是否已经存在相同的用户，则可以在 DaoUtils 类中新建 isExistUser 方法用于判断用户名是否存在（登录时也需要验证密码是否正确）。具体代码如下：

```
//判断用户名是否存在
public Boolean isExistUser (String[] values,Boolean isLogin){
    int count=0;
    database = dbHelper.getWritableDatabase();
    String sql="select * from t_user where   name=?";
    //登录时需要验证密码
    if(isLogin)
        sql+=" and password=?";
    Cursor cursor=getResultBySql(sql,values);
    count=cursor.getCount();
    cursor.close();
    closedDB();
    return count==0?false:true;
}
```

（2）在 RegisterActivity 类的 onCreate 方法中重写注册按钮的点击监听事件，以判断用户是否存在，如果不存在则将数据插入到 t_user 表中用于存储用户信息；如果已经存在，则提示"注册失败，该账户已存在！请重新输入！"。具体代码如下：

```
btn_qr.setOnClickListener(new View.OnClickListener() {
    @Override
    public void onClick(View view) {
        /*点击"确定"按钮后，判断密码与确认密码是否一致及带星号的输入项（用户名、密码、确认密码）是否不为空，若判断为真，则弹出悬浮窗，提示信息"恭喜！注册成功！"并结束当前界面（返回用户登录界面），否则所有输入框置空并弹出悬浮窗，提示信息"输入有误，请重新输入！"*/
        if(edt_pwd.getText().toString().trim().equals(edt_pwd1.getText().toString().trim())
                && !edt_usr.getText().toString().trim().equals("") ){
            radio_sex = (RadioButton)findViewById(radioGroup.getCheckedRadioButtonId());

            ContentValues values = new ContentValues();
            //判断用户名是否存在
```

```
                    Boolean isExistUser=daoUtils.isExistUser(new String[]{edt_usr.getText().toString()},false);
                    if(!isExistUser) {
                        values.put("name", edt_usr.getText().toString());
                        values.put("password", edt_pwd.getText().toString());
                        values.put("gender", radio_sex.getText().toString());
                        values.put("nick", edt_usr.getText().toString());
                        values.put("phone", edt_phone.getText().toString().trim());
                        values.put("email", edt_email.getText().toString().trim());
                        int count = daoUtils.insert("t_user", values);
                        if(count==1){
                            Toast.makeText(RegisterActivity.this,"恭喜注册用户写入数据库成功！"+edt_usr.getText().toString(), Toast.LENGTH_SHORT).show();
                            Intent intent=new Intent(RegisterActivity.this, LoginActivity.class);
                            startActivity(intent);
                        }else{
                            Toast.makeText(RegisterActivity.this,"注册失败，请重新输入！"+edt_usr.getText().toString(), Toast.LENGTH_SHORT).show();
                        }
                    }else{
                        Toast.makeText(RegisterActivity.this,"注册失败，该账户已存在！请重新输入！"+edt_usr.getText().toString(), Toast.LENGTH_SHORT).show();
                    }
                }else {
                    Toast.makeText(RegisterActivity.this,"输入有误，请重新输入！", Toast.LENGTH_SHORT).show();
                }
            }
        });
```

9.2 添加商品到购物车

实现添加商品到购物车的主要步骤如下。

（1）在购物商城 APP 中，点击某商品可以进入显示商品的详细信息页面。

（2）在该页面中，点击"加入购物车"按钮，将弹出确认添加到购物车的窗口。

（3）在该窗口中可以修改商品的购买数量，点击"确认"按钮，即可将该商品添加到购物车中，并关闭窗口，返回商品详细信息页面。

（4）在商品详细信息页面，点击"购物车"即可显示购物车中的商品。

（5）在购物车页面点击右上角的"编辑"按钮，可以增加或减少购物车中的商品数量。

（6）选中要购买的商品，点击"去结算"按钮，将弹出扫码支付窗口，用户可以进行扫码支付。

9.2.1 显示商品详细信息

在购物商城 APP 中，点击某件商品可以进入显示商品的详细信息页面。

（1）在分类模块 CategoryFragment 类中，重写右侧商品列表的点击事件的 onItemClick 方法，创建跳转页面的意图，并传递相应参数到商品详情页 GoodsDeatilsActivity 中。具体代码如下：

```
/**
 * 点击事件
 * @param itemView
```

```
 * @param position
 */
@Override
public void onItemClick(View itemView, int position) {
    Toast.makeText(getActivity().getApplicationContext(), "切换至商品详情", Toast.LENGTH_SHORT).show();
    //创建跳转页面的意图
    Intent startTo = activity(GoodsDetailsActivity.class);
    //传递序列化的对象
    startTo.putExtra(StaticUtils.SHOPID, commodityList.get(position).getId())
            .putExtra(StaticUtils.SHOPIMAGE, commodityList.get(position).getImageUrls());
    //跳转页面
    startActivityForResult(startTo, 1);
}
```

（2）创建商品详细信息页面的 Activity，并命名为 GoodsDetailsActivity，首先使其继承自 APPCompatActivity 类，并且实现 ViewPager 类的 OnPageChangeListener 接口、AddGoodsDialog 类的 OnGoodsChangeListener 接口和 View 类的 OnClickListener 接口，声明和初始化所需的变量，重写相关方法。关键代码如下：

```
/**
 * 商品详情界面
 */

public class GoodsDetailsActivity extends APPCompatActivity
        implements ViewPager.OnPageChangeListener,
        AddGoodsDialog.OnGoodsChangeListener,
        View.OnClickListener {
    protected DefaultTitleBar titleBar; //标题栏
    private String[] imageUrls; //图片的数组
    private ViewPager images; //展示图片
    private TextView imagePage; //显示页码
    private TextView title; //标题
    private TextView prompt; //提示
    private TextView price; //价格
    private TextView brand; //显示的品牌
    private TextView merchant; //显示的店铺
    private CheckBox focusOn; //关注
    private GoodDetails goodDetails; //商品详情
    Result<GoodDetails> result=new Result<GoodDetails>() ;
    private DaoUtils daoUtils; //数据库操作类
    private    MediaPlayer mediaPlayer; //播放背景音乐操作类
```

（3）初始化控件、添加监听及设置功能。关键代码如下：

```
/**
 * 初始化控件
 */

public void initView() {
    images = (ViewPager) findViewById(R.id.show_pictures);
    imagePage = (TextView) findViewById(R.id.show_page);
    title = (TextView) findViewById(R.id.show_title);
    prompt = (TextView) findViewById(R.id.show_prompt);
    price = (TextView) findViewById(R.id.show_price);
    brand = (TextView) findViewById(R.id.show_brand);
    merchant = (TextView) findViewById(R.id.show_merchant);
    focusOn = (CheckBox) findViewById(R.id.focus_on);
    shoppingCart = (CheckBox) findViewById(R.id.shopping_cart);
    addThing = (Button) findViewById(R.id.add_thing);

    badgeView = new BadgeView(this);
```

```java
        addGoodsDialog = new AddGoodsDialog(this);
    }
    /**
     * 添加监听
     */
    public void setListener() {
        images.addOnPageChangeListener(this); //滑动监听
        focusOn.setOnClickListener(this); //点击监听
        addThing.setOnClickListener(this); //点击监听
        addGoodsDialog.setOnGoodsChangeListener(this); //数量发生改变
        shoppingCart.setOnClickListener(this); //点击监听
    }
    /**
     * 设置功能，获取意图里的商品ID
     */
    public void setFunction() {
        if (null == titleBar){
            titleBar = new DefaultTitleBar(this);
        }
        daoUtils = new DaoUtils(this); //初始化数据库操作类
        titleBar.setTitle("商品详情");
        //获取意图里的商品ID
        Intent intent = getIntent(); //获取意图
        String shopID = intent.getStringExtra(StaticUtils.SHOPID); //商品ID
        String shopImage = intent.getStringExtra(StaticUtils.SHOPIMAGE); //图片的地址
        //根据商品ID加载商品详情
        downLoadDataFromNet(shopID); //获取指定的商品详情
        imageUrls = shopImage.split(StaticUtils.THESEPARATOR);
        images.setAdapter(new ImageAdapter(imageUrls, this)); //绑定适配器
        imagePage.setText(showPage(1)); //设置默认显示的页数
    }
```

（4）创建在设置功能的 setFunction 方法中用到的 downLoadDataFromNet 方法，根据商品 ID 查询数据库、获取商品详情对象、发送消息队列、更新页面。关键代码如下：

```java
/**
 * 根据ID查询商品信息，发送消息队列更新页面
 */
private void downLoadDataFromNet(String id) {
    //根据ID查询商品信息
    result=daoUtils.queryResultGoodDetailsById(id);
    //发送消息队列更新页面
    handler.obtainMessage( HttpCode.DETAILS,result).sendToTarget();
}
```

（5）初始化消息队列，根据消息类型将数据展示到视图上。关键代码如下：

```java
private Handler handler = new Handler(new Handler.Callback() {
    @Override
    public boolean handleMessage(Message msg) {
        switch (msg.what) {
            case HttpCode.DETAILS://详情
                //根据消息类型判断是否更新视图数据
                if (TextUtils.equals(HttpCode.SUCCESS, result.getReason())) {
                    goodDetails = result.getResult();
                    addDataToView(); //将数据展示到视图上
                } else {
                    Toast.makeText(GoodsDetailsActivity.this.getApplicationContext(), "加载数据失败！", Toast.LENGTH_SHORT).show();
```

```
                    }
                    break;
                case HttpCode.ERROR://错误
                    Toast.makeText(GoodsDetailsActivity.this.getApplicationContext(), "加载数据失败！", Toast.LENGTH_SHORT).show();
                    break;
            }
            return false;
        }
    });
```

（6）将查询的数据展示到指定的控件上。关键代码如下：

```
/**
 * 将查询的数据展示到指定的控件上
 */
private void addDataToView() {
    GetViewTextUtils.setText(title, goodDetails.getTitle()); //设置标题
    String promptStr = goodDetails.getPrompt(); //获取提示文字
    //不为空就展示
    if (!TextUtils.isEmpty(promptStr)) {
        GetViewTextUtils.setText(prompt, promptStr); //设置提示文字
    } else {
        prompt.setVisibility(View.GONE);
    }
    //设置显示的价格
    price.setText(FormatUtils.getPriceText(goodDetails.getNowPrice()));
    //设置显示的品牌
    brand.setText(StaticUtils.BRAND + goodDetails.getBrand());
    //设置显示的店铺
    merchant.setText(StaticUtils.MERCHANT + goodDetails.getMerchant());
}
```

（7）运行该项目，效果如图 9.1 所示。

图 9.1　商品详情页面运行效果

9.2.2 将商品加入购物车

在商品详情页面点击"加入购物车"按钮,即可将该商品添加到购物车中,具体实现步骤如下:

(1)添加"加入购物车"按钮点击事件监听器,在重写 onClick 方法时用 switch 语句处理"加入购物车"的情况。这里主要演示添加功能,关键代码如下:

```java
/**
 * 点击事件
 *
 * @param v
 */
@Override
public void onClick(View v) {
    switch (v.getId()) {
        case R.id.add_thing://添加
            addGoodsDialog.setGoodDetails(goodDetails);
//根据图片名称获取图片 ID,再根据图片 ID 获取图片路径,并设置 addGoodsDialogD 的图片 url
            addGoodsDialog.setFirstImageUrl(FileUtils.getResourcesUri(this.getResources(),FileUtils.getDrawableId( imageUrls[0])
            );
            addGoodsDialog.show();
            break;
    }
}
```

(2)创建 AddGoodsDialog 加入购物车对话框,继承自 CustomDialog 类,并且实现 ImageDownLoadListener 类和 View 类的 OnClickListener 接口,声明并初始化所需变量,然后重写对应方法,关键代码如下:

```java
public class AddGoodsDialog extends CustomDialog implements ImageDownLoadListener, View.OnClickListener {
    private GoodDetails goodDetails;//商品详情
    private String firstImageUrl;//第一张图片的网络连接
    private ImageView image;//展示图片
    private Button decision;//确定
    private TextView title;//标题
    private TextView brand;//品牌
    private TextView merchant;//品牌
    private TextView price;//价格
    private TextView id;//商品 ID
    //选择商品的数量
    private ImageButton addCount;//增加数量
    private ImageButton cutCount;//减少数量
    private EditText inputCount;//输入数量
    private int selectGoodsCounts;//选择的商品的数量
    private String imageSavePath;//图片保存的路径
    private Bitmap resultBitmap;//返回的图片
    private DaoUtils daoUtils;//操作数据库
}
```

（3）重写 onClick 方法，添加"确认"按钮点击事件。点击"确认"按钮首先需要将图片添加到手机上，然后调用 addSelectCountToTable 方法将数据保存到数据库，同时判断数据是否存在，如果存在就更新数据，如果不存在就添加数据。关键代码如下：

```
ContentValues values = new ContentValues();
values.put("_id", goodDetails.getId());//商品 ID 编号
values.put("title", goodDetails.getTitle());//标题
values.put("price", goodDetails.getNowPrice());//价格
values.put("brand", goodDetails.getBrand());//品牌
values.put("image_url", goodDetails.getImageUrls());//图片的网址
values.put("image", imageSavePath);//图片
values.put("merchant", goodDetails.getMerchant());//商家
boolean dataChange;
if (haveData) {//修改数据库中的数据
    Cursor cursor = daoUtils.getResultCursor(StaticUtils.CART_TABLE,
            new String[]{"count"},"_id = ?",new String[]{goodDetails.getId()},null);
    cursor.moveToFirst();//将游标移动到最开始
    int count = cursor.getInt(cursor.getColumnIndex("count"));
    cursor.close();//关闭游标
    daoUtils.closedDB();//关闭数据库的连接
    values.put("count", selectGoodsCounts + count);//数量
    dataChange = daoUtils.update(StaticUtils.CART_TABLE, values, "_id = ?",
            new String[]{goodDetails.getId()})
            == DBStateCode.OPERATION_SUCCESS ? true : false;
} else {//向数据库添加新的数据
    values.put("count", selectGoodsCounts);//数量
    dataChange = daoUtils.insert(StaticUtils.CART_TABLE, values)
            == DBStateCode.OPERATION_SUCCESS ? true : false;
}
return dataChange;
```

（4）添加购物车页面主要布局 dialog_add_goods.xml，如图 9.2 所示。

图 9.2 添加购物车页面主要布局

150

（5）运行该代码，效果如图 9.3 所示。

图 9.3　将商品添入购物车效果

9.2.3　查看、编辑购物车

用户将商品添加到购物车后，可以在商品详情页点击"购物车"按钮，查看购物车中的商品列表，具体实现步骤如下。

（1）重写在 GoodDetailsActivity 类的 onClick 方法，添加 switch 语句的一个 case 子句，用于处理查看购物车事件。主要功能是将页面跳转到 ShoppingCartActivity 页面，关键代码如下：

```
/**
 * 点击事件
 *
 * @param v
 */
@Override
public void onClick(View v) {
    switch (v.getId()) {
        case R.id.focus_on://关注
            Toast.makeText(GoodsDetailsActivity.this, "关注功能即将开放，敬请期待！", Toast.LENGTH_SHORT).show();
            break;
        case R.id.shopping_cart://购物车
            Toast.makeText(GoodsDetailsActivity.this, "跳转购物车！", Toast.LENGTH_SHORT).show();
            Intent intent = new Intent(GoodsDetailsActivity.this, ShoppingCartActivity.class);
            startActivity(intent);//购物车页面
            closeMusic();
            finish();//关闭当前页面
            break;
        case R.id.add_thing://添加
            addGoodsDialog.setGoodDetails(goodDetails);
            addGoodsDialog.setFirstImageUrl(FileUtils.getResourcesUri(this.getResources(),FileUtils.getDrawableId( imageUrls[0])));
            addGoodsDialog.show();
```

```
            break;
    }
}
```

(2) 创建 ShoppingCartActivity 类，使其继承自 APPCompatActivity 类，然后指定布局文件，设置功能加载购物车页面，关键代码如下：

```java
import android.os.Bundle;
import android.support.v7.APP.APPCompatActivity;

import com.mingrisoft.mrshop.mrshop.R;

import com.mingrisoft.mrshop.fragment.ShoppingCartFragment;

/**
 * 购物车页面
 */
public class ShoppingCartActivity extends APPCompatActivity {
    @Override
    protected void onCreate(Bundle savedInstanceState) {
        super.onCreate(savedInstanceState);
        setContentView(R.layout.activity_shopping_cart);
        setFunction();
    }
    /**
     * 设置功能
     */
    public void setFunction() {
        ShoppingCartFragment cartFragment = new ShoppingCartFragment();//实例化购物车的Fragment对象
        cartFragment.setBackBtnShow(true);      //显示返回按钮
        //加载购物车页面
        getSupportFragmentManager()
                .beginTransaction()
                .add(R.id.activity_shopping_cart,cartFragment)
                .commit();
    }
}
```

(3) 创建购物车的 Fragment 对象，名称为 ShoppingCartFragment，使其继承自 TitleFragment 类，并实现 ExpandableListVicw 类的 OnGroupClickListener 接口、ShopCartListener 接口、View 类的 OnClickListener 接口，以及 ExpandableListView 类的 OnChildClickListener 接口，然后声明和初始化所需变量，并且重写初始化、设置功能等相关方法，关键代码如下：

```java
public class ShoppingCartFragment extends TitleFragment
        implements ExpandableListView.OnGroupClickListener,
        ShopCartListener, View.OnClickListener,ExpandableListView.OnChildClickListener {

    private boolean titleBarIsShow; //显示标题栏
    private boolean isInit = true; //是否是初始化
    private DaoUtils daoUtils; //数据库操作类
    private List<GoodsShop> carts; //购物车数据
    private ExpandableListView listView; //购物车列表
    private TextView allPrice; //总价格
    private TextView allCount; //总数量
    private ShopCartAdapter shopCartAdapter; //购物车列表适配器
    private CheckBox selectAll; //选择全部
    private RelativeLayout empty; //空
```

```
    private TextView rightView; //右侧的编辑按钮
    private Button cartCommit; //提交订单
    private LinearLayout setBarOne; //设置栏1
    private LinearLayout setBarTwo; //设置栏2
    private Button share, focus, delete; //分享、关注、删除
    private String saveStr, oldStr; //保存数据
        private LocalBroadcastManager localBroadcastManager; //本地广播管理器
```

（4）设置视图布局，初始化控件，设置监听，设置功能，关键代码如下：

```
/**
 * 设置视图的布局
 *
 * @return
 */
@Override
protected int setLayoutID() {
    return R.layout.fragment_shoppingcart;
}

/**
 * 初始化控件
 *
 * @param view
 */
@Override
public void initView(View view) {
    listView = (ExpandableListView) view.findViewById(R.id.cart_list);
    allPrice = (TextView) view.findViewById(R.id.all_price);
    allCount = (TextView) view.findViewById(R.id.all_count);
    selectAll = (CheckBox) view.findViewById(R.id.cart_all);
    empty = (RelativeLayout) view.findViewById(R.id.empty);
    setBarOne = (LinearLayout) view.findViewById(R.id.set_bar_1);
    setBarTwo = (LinearLayout) view.findViewById(R.id.set_bar_2);
    cartCommit = (Button) view.findViewById(R.id.cart_commit);
    share = (Button) view.findViewById(R.id.cart_share);
    focus = (Button) view.findViewById(R.id.cart_shop_focus);
    delete = (Button) view.findViewById(R.id.cart_delete);
}
/**
 * 设置监听
 */
@Override
public void setListener() {
    listView.setOnGroupClickListener(this);
    listView.setOnChildClickListener(this);
    selectAll.setOnClickListener(this);
    cartCommit.setOnClickListener(this);
    share.setOnClickListener(this);
    delete.setOnClickListener(this);
    focus.setOnClickListener(this);
}
/**
 * 设置功能
 */
@Override
public void setFunction() {
    titleView.setTitle("购物车");
    titleView.backBtnIsShow(titleBarIsShow);
    titleView.setTitleBarBackgroundColor(Color.WHITE);
```

```java
        titleView.setTitleColor(Color.BLACK);
        titleView.setBackBtnIcon(R.drawable.title_back_btn2);
        addTitleRightView();//添加标题栏右侧的按钮
        //添加数据为空时展示的页面效果视图
        listView.setEmptyView(empty);
        daoUtils = new DaoUtils(mContext);//初始化数据库操作类
        if (isInit) {//第一次加载页面
            addDataToView();//将数据展示到页面中
            isInit = false;//设置为false
        }
    localBroadcastManager = LocalBroadcastManager.getInstance(getActivity());
```

（5）加载页面，将数据展示到页面中，关键代码如下：

```java
/**
 * 将数据展示到页面中
 */
private void addDataToView() {
    showBottomBar(false);
    isMake = false;
    setRightTextViewShow(StaticUtils.EDITOR, R.color.default_text_color);
    //加载数据到购物车列表中
    List<GoodsCart> list = daoUtils.getResultListAll(StaticUtils.CART_TABLE, GoodsCart.class);
    initCartData(list);//初始化购物车数据
    shopCartAdapter = new ShopCartAdapter(carts, mContext);
    listView.setAdapter(shopCartAdapter);
    for (int i = 0; i < carts.size(); i++) {//默认全部展开
        listView.expandGroup(i);
    }
    shopCartAdapter.setShopCartListener(this);
    shopCartAdapter.getCustomSelectData();
    rightView.setOnClickListener(new View.OnClickListener() {
        @Override
        public void onClick(View v) {
            if (isMake) {//完成
                initCartData(daoUtils.getResultListAll(StaticUtils.CART_TABLE, GoodsCart.class));
                makeNewSaveJsonMessage(); //验证修改后的数据生成新的保存数据
                setEditorEndShow(); // 设置修改之后的显示效果
                setRightTextViewShow(StaticUtils.EDITOR, R.color.default_text_color);
            } else {//编辑
                saveStr = new Gson().toJson(carts); //保存数据为JSON
                selectAll.setChecked(false); //解除全选按钮的选中状态
                shopCartAdapter.changAllDataSelectState(false);
                isMake = true;
                setRightTextViewShow(StaticUtils.COMPLETE, R.color.black);
            }
            showBottomBar(isMake);
        }
    });
}
```

（6）初始化购物车中的数据，关键代码如下：

```java
/**
 * 初始化购物车数据
 *
 * @param list
 */
private void initCartData(List<GoodsCart> list) {
    if (null == carts) {
```

```
        carts = new ArrayList<>();//商品类型数据结合
    } else {
        carts.clear();//清空数据集合
    }
    Cursor cursor = daoUtils.getResultCursor(StaticUtils.CART_TABLE,
            new String[]{"merchant"}, null, null, "merchant", null, null);//分组查询
    while (cursor.moveToNext()) {
        GoodsShop goodsShop = new GoodsShop();
        goodsShop.setMerchant(cursor.getString(cursor.getColumnIndex("merchant")));
        goodsShop.setCartsList(new ArrayList<GoodsCart>());//初始化集合数据
        carts.add(goodsShop);//添加购物车商品类型
    }
    cursor.close();//关闭游标
    daoUtils.closedDB();//关闭数据库连接
    for (GoodsCart goodsCart : list) {//循环处理数据
        String shopName = goodsCart.getMerchant();//获取店铺名称
        int shopCount = goodsCart.getCount();//获取选中的数量
        goodsCart.setViewState(new CartViewState(
                true, shopCount > 1 ? true : false, false));//初始化控件的状态
        int position = getIndex(shopName);//索引值
        if (position == -1) {
            return;//等于-1 说明出现问题
        } else {//将指定的数据添加到指定的数据集合中
            carts.get(position).getCartsList().add(goodsCart);//添加数据
        }
    }
    oldStr = new Gson().toJson(carts);//保存数据
}
```

（7）运行该项目，效果如图 9.4 所示。

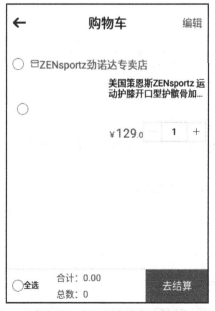

图 9.4　购物车页面运行效果

9.2.4 购物车结算

在购物车页面中选中要购买的商品后,可以点击"去结算"按钮进行结算,具体实现步骤如下。

(1)在 weight 包中创建支付窗口类,名为 PayDialog,使其继承自 CustomDialog 类,并实现 View 类的 OnclickListener 接口,然后重写相关方法,关键代码如下:

```java
/**
 * 功能:支付窗口
 */
public class PayDialog extends CustomDialog implements View.OnClickListener{
    private Button btn_pay;    //完成支付按钮
    public PayDialog(Context context) {
        super(context,true);
    }
    @Override
    protected void onCreateView(WindowManager windowManager) {
        DisplayMetrics outMetrics = new DisplayMetrics();
        windowManager.getDefaultDisplay().getMetrics(outMetrics);
        setDialogWidth(WindowManager.LayoutParams.MATCH_PARENT);
        setDialogHeight(WindowManager.LayoutParams.WRAP_CONTENT);
        setDialogGravity(Gravity.BOTTOM);
        setContentView(R.layout.dialog_pay);    //设置布局文件
        findId();  //绑定控件
    }
    private   void findId(){
        btn_pay = (Button) findViewById(R.id.mr_finish);
        btn_pay.setOnClickListener(this); //设置监听器
    }

    @Override
    public void onClick(View v) {
        switch (v.getId()) {
            case R.id.mr_finish://完成按钮,模拟支付成功
                dismiss();    //关闭窗口
                break;
        }
    }
}
```

(2)在 ShoppingCartFragment 类中声明 PayDiaLog 类对象,并且在 initView 方法中创建支付窗口,关键代码如下:

```java
private PayDialog payDialog;//添加支付窗口
/**
 * 初始化控件
 *
 * @param view
 */
@Override
public void initView(View view) {
    …
    payDialog = new PayDialog(view.getContext());        //创建支付窗口
}
```

(3)在 ShoppingCartFragment 类 onClick 方法的 switch 语句中,添加一个判断语句,当购

物车里添加了待购买的商品，则显示支付窗口，关键代码如下：

```
case R.id.cart_commit://结算
    if(mCarts.size() > 0){
        payDialog.show();          //显示支付窗口
    }else{
        Toast.makeText(getContext(), "请先选择要购买商品！", Toast.LENGTH_SHORT).show();
        return;
    }
    break;
```

（4）在 ShoppingCartFragment 类的 setFunction 方法（设置功能方法）中添加支付窗口，关闭监听器，用于实现支付窗口关闭后清空购物车并返回商城首页，关键代码如下：

```
//支付窗口关闭时
payDialog.setOnDismissListener(new DialogInterface.OnDismissListener() {
    @Override
    public void onDismiss(DialogInterface dialog) {
        if(getActivity().getLocalClassName().equals("activity.MainActivity")) {
            MainActivity activity = (MainActivity) getActivity();
            activity.radioGroup.check(R.id.mr_shoppingmall);   //切换到首页
        }else   if(getActivity().getLocalClassName()
                .equals("activity.ShoppingCartActivity")){
            Intent intent=new Intent(getActivity(),MainActivity.class);
            intent.setFlags(Intent.FLAG_ACTIVITY_CLEAR_TOP);
            getActivity().startActivity(intent);      //购物车主页面
        }
        Intent intent = new Intent(StaticUtils.THING_COUNT)//发送广播的意图，并更新购物车数量标记
                .putExtra(StaticUtils.NOW_COUNT, 0);
        localBroadcastManager.sendBroadcast(intent); //发送广播
        daoUtils.deleteAll(StaticUtils.CART_TABLE);          //清空购物车
    }
});
```

（5）运行该代码，效果如图 9.5 所示。

图 9.5　购物车结算运行效果

附录 A 素材说明

购物商城 APP 完整项目代码的组织结构和说明如附图 1 所示。

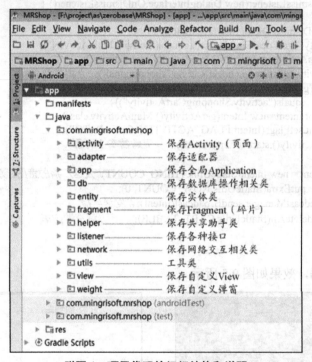

附图 1 项目代码的组织结构和说明

代码目录包说明表如附表 1 至附表 10 所示。

附表 1 activity 包说明

类　名	功　能
GoodsDetailsActivity	商品详情页面
LoginActivity	登录页面
MainActivity	首页页面
RegisterActivity	注册页面
ShoppingCartActivity	购物车页面
ShopTypeActivity	分类模块页面

附表 2　adapter 包说明

类　　名	功　　能
base	适配器基础包
listener	点击事件监听
CommodityAdapter	商城列表适配器
ImageAdapter	展示图片适配器
ShopCartAdapter	购物车数据适配器
ShopCartListener	购物车监听器
TypeAdapter	类型适配器

附表 3　entity 包说明

类　　名	功　　能
Result	返回实体类
CartViewState	购物车中控件的状态对象
Commodity	商品对象
GoodDetails	商品详情对象
GoodsCart	购物车对象
GoodsShop	购物车的商品类型对象
HomeEntity	首页的数据对象
ShapeType	选中状态对象
User	用户对象

附表 4　fragment 包说明

类　　名	功　　能
BaseFragment	Fragment 基类
TitleFragment	带有标题栏的 Fragment
CategoryFragment	分类（主要实现双列表联动）
HomeFragment	首页
MineFragment	个人中心
ShoppingCartFragment	购物车

附表 5　listener 包说明

类　　名	功　　能
ImageDownLoadListener	获取图片监听

附表 6　service 包说明

类　　名	功　　能
MusicServer	商品详情播放音乐服务

附表 7 utils 包说明

类 名	功 能
DensityUtils	屏幕像素密度转换工具类
DownLoadImageUtils	加载图片工具类
FileUtils	文件操作类
FormatUtils	格式化工具类
GetViewTextUtils	文本内容工具类
HttpClient	请求对象工具类
HttpUtils	网络工具类
StaticUtils	常量工具类

附表 8 view 包说明

类 名	功 能
CustomScrollListener	滑动监听
DefaultTitleBar	默认的标题栏
LoadMoreListener	加载更多监听
LoadMoreView	加载更多布局效果

附表 9 dialog 包说明

类 名	功 能
AddGoodsDialog	添加购物车对话框
ChangeDialog	购物车对话框
CustomDialog	购物车对话框
MineDialog	个人中心对话框
PayDialog	支付对话框

附表 10 layout 包说明

类 名	功 能
activity_goods_details.xml	商品详情主页面
activity_goods_details_body.xml	商品详情中间页面
activity_goods_details_select_bar.xml	商品详情底部页面
activity_login.xml	登录页面
activity_main.xml	首页页面
activity_register.xml	注册页面
activity_shop_type.xml	商品类型页面
activity_shopping_cart.xml	购物车主页面
bottombar_layout.xml	首页底部切换页面
default_title_bar_.xml	默认标题栏页面
dialog_add_goods.xml	添加商品对话框页面
dialog_mine.xml	个人中心对话框页面
dialog_pay.xml	支付对话框页面
fragment_category.xml	分类页面

续表

类　　名	功　　能
fragment_home.xml	首页中间分类导航页面
fragment_mine.xml	个人中心页面
fragment_shoppingcart.xml	购物车页面
item_cart_child.xml	购物车列表页面
item_cart_group.xml	购物车店铺页面
item_commodity.xml	商品详情页面
item_commodity2.xml	商品详情页面
item_image.xml	图片页面
item_type_name.xml	分类页面
view_body_change_count.xml	购物数量页面
view_empty_layout.xml	购物车无数据页面
view_load.xml	加载提示页面
view_load_more.xml	加载更多提示页面
view_none.xml	加载没有更多提示页面
view_select_count.xml	数量选择页面

华信SPOC官方公众号

欢迎广大院校师生 **免费** 注册应用

www.hxspoc.cn

华信SPOC在线学习平台

专注教学

- 数百门精品课 数万种教学资源
- 教学课件 师生实时同步
- 电脑端和手机端（微信）使用
- 多种在线工具 轻松翻转课堂
- 一键引用，快捷开课 自主上传，个性建课
- 测试、讨论、投票、弹幕…… 互动手段多样
- 教学数据全记录 专业分析，便捷导出

登录 www.hxspoc.cn 检索 华信SPOC 使用教程 获取更多

华信SPOC宣传片

教学服务QQ群： 1042940196
教学服务电话： 010-88254578/010-88254481
教学服务邮箱： hxspoc@phei.com.cn

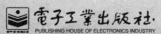

电子工业出版社　　　华信教育研究所
PUBLISHING HOUSE OF ELECTRONICS INDUSTRY